KASHMIR

KASHMIR:

ITS NEW SILK INDUSTRY,

WITH SOME ACCOUNT OF ITS
NATURAL HISTORY, GEOLOGY, SPORT, ETC.

SIR THOMAS WARDLE

GYAN

Published by

Gyan Publishing House
5, Ansari Road
Daryaganj, New Delhi-110002
Phone: 011-47034999, 9811692060
E-mail: books@gyanbooks.com

Distribution Network
gyanbooks.com
India, USA, Canada, UK, Australia, France

ISBN: 978-81-212-8740-1 (PB)
First Published, 1904

2nd Impression 2023

Printed at: Gyan Press, Delhi.

KASHMIR
Author: Thomas Wardle

Thomas Wardle

KASHMIR:

ITS NEW SILK INDUSTRY,

WITH SOME ACCOUNT OF ITS

NATURAL HISTORY, GEOLOGY, SPORT, ETC.,

AND WITH
FORTY-FIVE FULL-PLATE ILLUSTRATIONS OF KASHMIR
SCENERY, AND OF SPORT, INDUSTRY, FOSSILS,
ETC., ALSO
NOTES OF A VISIT TO THE SILK-PRODUCING
DISTRICTS OF BENGAL IN 1885-6,

BY

SIR THOMAS WARDLE,

J.P., F.G.S., F.C.S.

Chevalier de la Legion d'Honneur, and Officier d'Academie of France; Member of the Silk Juries of the Paris Exhibitions of 1878 and 1889; Honorary Superintendent, Silk Culture Court, Indian and Colonial Exhibition, London, 1886; Chairman, Silk Section, Manchester Royal Jubilee Exhibition, 1887; late Examiner on Silk and Silk Dyeing, City and Guilds of London Institute; President of the Silk Association of Great Britain and Ireland, Hon. Sec., Ladies' National Silk Association, etc.

LONDON:

SIMPKIN, MARSHALL, HAMILTON, KENT AND CO.

LEEK:

W. H. EATON, THE MOORLANDS PRESS, DERBY STREET.

1904.

PUBLICATIONS BY THE AUTHOR.

GEOLOGY OF LEEK AND OF CROMER.

ON CARBONIFEROUS LIMESTONE AND ITS MICROSCOPIC CRYSTALS OF SILICA.

THE WILD SILKS OF INDIA.

RESEARCHES ON SILK FIBRE.

ROYAL COMMISSION REPORT ON THE ENGLISH SILK INDUSTRY, 1885.

DESCRIPTIVE SILK CATALOGUE; COLONIAL AND INDIAN EXHIBITION, 1886.

ON SILK. MANCHESTER JUBILEE EXHIBITION, 1887.

ECONOMIC UTILIZATIONS OF THE WILD SILKS OF INDIA.

EARLY COPTIC TIMES, CHURCHES, FABRICS AND DYES.

SILK; ITS ENTOMOLOGY, HISTORY AND MANUFACTURE.

ON 630 SPECIES OF SILK-PRODUCING LEPIDOPTERA.

NEW FACTS IN THE STAFFORDSHIRE YOREDALE MEASURES.

ON SEWAGE TREATMENT AND DISPOSAL.

ON SILK POWER-LOOM WEAVING IN LYONS.

ON CYPRUS SILKS AND TANS.

THE "BREAKING" OF COPMERE.

TEXTILE PRINTING AS AN ART.

ON CONTINENTAL ADULTERATION OF SILK BY CHEMICAL WEIGHTING.

ETC., ETC,

DEDICATED

BY GRACIOUS PERMISSION

TO

HER ROYAL HIGHNESS THE PRINCESS

OF WALES,

PRESIDENT OF THE LADIES' NATIONAL

SILK ASSOCIATION,

WITH PROFOUND RESPECT.

PREFACE.

IN order to make this book of greater practical value I have inserted my various Reports just as they were written and sent to the Government of India and to the State of Kashmir. They give an accurate account of the Sericicultural operations in Kashmir, and contain a large amount of technical and practical information which cannot fail to be of service to those in other parts of India and the Colonies desiring to establish and conduct Sericicultural operations successfully.

These Reports are also a record of my work in and for Kashmir during the last seven years, and which has fortunately proved both successful and encouraging in the establishment of a new Kashmir industry.

My object in reprinting the Report of the Conference held at Calcutta in 1886, under the Chairmanship of Sir Edward C. Buck, then Secretary of Revenue and Agriculture, is to re-urge consideration of the suggestions I then made, and to draw attention to the information contained in that Report, both with regard to Bengal Sericiculture, and to the more energetic production of the Wild Silks of India, chiefly that of Tussur Silk, now that its manufacture in Europe and America has so very largely increased, and the extensive demand for it is permanently established.

Nowhere is scientific Sericiculture more required than in the Tussur districts of the Central Provinces, where the Tussur industry has seriously declined. The cocoons have become much smaller from want of breeding knowledge, and the proportion of deaths from disease is so great that the industry is hardly worth following.

In the neighbouring districts of Bengal, with identical climatic conditions, the Tussur cocoons are about three times larger and failure from disease is rare.

The natives should be allowed to go into the forest to collect suitable cocoons for breeding. I think it is important that the Forest Department should open the forests of the Central Provinces to them for this purpose.

I further earnestly repeat my conviction that a Central Imperial Sericicultural Station in India, like those of Montpellier in France and Padua in Italy, with branches in the silk-producing districts, as in Italy, as may gradually be required, would be of immense usefulness in India in preventing silkworm disease, in teaching microscopic manipulation, thereby hastening the time when eggs of reproduction could be bred with safety (an impossibility now and for some time to come); besides many other important advantages which would accrue to industrial Sericiculture when conducted upon the scientific basis recorded in Chapter XVII.

I dare say the latter portion of my book will be read with interest by those who are fond of natural history and sport. I had a few fortunate and exceptional opportunities for both; and short as the time at my disposal was, the experience I was privileged to have was of a very charming description, particularly that part of it spent geologically at considerable altitudes in the Himalayas.

The Princess of Wales has graciously permitted me to dedicate my book to Her Royal Highness.

The Princess has for several years been taking the greatest interest in the Silk Industries of the Empire, and is President of the Ladies' National Silk Association, which now numbers several thousand members, and has a Council of ninety of the principal ladies of the country. Its object is the promotion of the national Silk Industry.

Her Royal Highness the late Duchess of Teck was the first President of the Association, and devoted much time and energy to its well-being.

My thanks are due to Geoffrey Millais, Esq., for so kindly allowing me to reproduce three of his photographs.

Also to my niece, Miss Elinor Wardle, for her interesting drawings of the fossils in plate 41.

I wish to acknowledge my indebtedness to Miss K. Lowe for her valuable assistance in helping me to revise and correct proofs.

LEEK, AUGUST, 1904.

CONTENTS.

LIST OF ILLUSTRATIONS.

PLATE | PAGE

CHAPTER I.

INTRODUCTORY, BEING MOTIVES FOR SERICICULTURAL ENQUIRY IN INDIA. EFFORTS TO UTILISE TUSSUR SILK. BLEACHING AND DYEING DISCOVERIES. TUSSUR SILK STATISTICS AND LARGELY INCREASING CONSUMPTION. KASHMIR SILK, ITS POTENTIALITIES.

I NOW beg to submit my Reports to the Government of India on my efforts during the last few years to promote Sericiculture in Kashmir.

Before doing so it may be desirable that I should describe the earlier history of the movement as far as I am concerned, and what led to my taking an interest in it.

I was first led into research connected with the industrial development of India by my efforts to utilize the wild Tussur silk* of that Country. I was fortunate in being the first to succeed in dyeing this silk properly.

The results of these efforts were, at the request of the then Prince of Wales (His present Majesty

* Also written Tasar, Tussah, Tusser, Tussur, and Tussore, the Indian word being tasar, probably derived from the Sanskrit *tasara*, "a shuttle," to which the cocoon has a resemblance. It is the most important species of the family Saturnidæ, Antheræa Mylitta.

the King-Emperor), communicated to me by Sir Phillip Cunliffe-Owen, placed in the Indian Section of the Paris Exhibition of 1878, and were afterwards removed to the South Kensington Museum. They consisted of hanks of bleached and dyed Tussur silk yarns, and of woven and printed specimens, including the first piece of plush made from Tussur silk, which suggested the later large Yorkshire Industry of "seal cloth," an imitation of sealskin, chiefly made from the waste of Tussur silk; the first to exploit it being Messrs. Field & Bottrill of Skelmanthorpe, near Huddersfield, after having seen the Tussur silk exhibit in the South Kensington Museum.

For this Exhibit a Gold Medal was awarded at Paris to the Secretary of State for India. It was first awarded to me, but the Exhibit being entirely of Indian Tussur Silk, I decided that it ought to go to the Secretary of State for India.

Shortly after the Paris Exhibition of 1878, I was requested by the Lyons Chamber of Commerce to exhibit in Lyons on their behalf the results of my researches in the bleaching, dyeing, and utilization of this silk; and I arranged a representative collection of the material, raw, dyed, and manufactured, which was supplied and presented by the Government of India, clearly showing the improvements in the bleaching and dyeing processes, as well as indicating its commercial possibilities of textile utilization. The collection was exhibited in the Musée de la Bourse at Lyons, and named the

"Wardle Collection," and it remained on exhibition there for several years.*

In a comprehensive survey of the sericiculture and native Indian manufacture of Tussur silk, entitled "The Indian Silkworm" by Sir George (then Dr.) Birdwood, made in his Annual Report on the Government Central Museum, Bombay, for the year 1859-60, will be found many interesting particulars on this subject.

This Report is reprinted at length in my Handbook of the Collection illustrative of the Wild Silks of India in the Indian Section of the South Kensington Museum, published by the Science and Art Department of the Committee of Council on Education, South Kensington, 1881.

Sir George, who afterwards from 1871 to 1899, became Special Assistant in the Revenue and Statistics Department of the India Office, in this Report reviews the whole literature, historical and natural-historical of Tussur silk up to that early period of its publication.

The Silk Jury of the Paris Exhibition of 1878 were not slow to see the importance of the improvements made in the bleaching and dyeing of this silk. The President of the Jury told me that it was destined to have a gradual and important

* The natural colour of the Tussur cocoon and of its reeled silk thread is of a dark fawn hue which had hitherto prevented the dyer from obtaining rich colours upon it, as well also that the fibre is physically more intractable to tinctorial treatment than that of the Bombyx Mori, the larvæ of which produce the silk of Commerce.

development in Lyons. In time a large demand arose which the natives of India were too apathetic to supply; and it seems very unfortunate that India should have allowed China to have taken the lead, and chiefly to furnish the ever increasing main export supply of Tussur silk. Considering that only in one week last year 142 bales of Chinese Tussur silk were conditioned in Lyons alone, it is now probably too late for India to attempt to monopolise this market; but she should take up this subject in earnest, and endeavour to supply a part, if not the whole of this expanding trade.

It was undoubtedly Sir George Birdwood who was the first to call attention to this silk, which he, at that early date, thought might be fitted for more refined manufacturing treatment than it had received in India and elsewhere.

Previously to 1878 no Tussur silk was successfully used in Europe, although it had been often tried.

The difficulties of dyeing and bleaching this wild silk having however been successfully overcome, it gradually from that time began to be used, and increasing quantities were passed regularly through the Lyons Conditioning House each week, and on the 30th November, 1899, no less a quantity than 433 bales or 60,620 lbs. were conditioned that week,—a record week, and in one week in June of the present year 238 bales were conditioned there.

In 1902, 4,795 bales, weighing 703,942 lbs., passed through the Lyons Conditioning House, and in 1903, 5,141 bales, or 792,926 lbs.

In addition to this, from June, 1902, to May, 1903, 142 bales or 18,744 lbs. of Tussur Silk were exported from Shanghai to London, and 2,467 bales or 325,644 lbs. to America; while from June, 1903, to November, 1903, 5 bales (660 lbs.) were exported to London, and 1,220 bales (161,040 lbs.) to America. For the week ending February 13th 142 bales were, as above stated, conditioned at Lyons.

My work in connection with Indian Tussur silk is recorded in my Monograph of the Wild Silk Industry of India, added as an appendix (26 pages) with micrometer measurements of silk fibres, to Sir George Birdwood's Handbook of the British-India Section of the Paris Universal Exhibition, 1878; also in my contribution of 38 pages in French to Monsieur Rondot's Work, "L'art de la Soie;" and in my Paper on Tussur silk, read at a meeting of the Society of Arts, held on the 14th May, 1891, presided over by the late Lady Egerton of Tatton; and also in my "Handbook on the Wild Silks of India" above mentioned.

In 1885 I was sent to India by the Government of India and the Royal Commission of the Colonial Exhibition with a two-fold object, one to report on sericiculture in Bengal, and the other, to collect typical silks from various parts of India for exhibition at the forthcoming Colonial and Indian Exhibition of 1886.

For this latter purpose I was requested to go by his present Majesty (then Prince of Wales) who was the President of the British Section of that Exhibition.

After visiting the Bengal silk-producing districts of Surdah and Berhampur, I was requested to give an account of my visit by Mr. (now Sir E. C.) Buck, then Secretary to the Government of India in their Revenue and Agricultural Department.

I did so at a public meeting held in Calcutta for the purpose on the 8th January, 1886, and my Reprt was published as a Blue-book at Calcutta, and duly forwarded to the India Office. See Chapter 18.

I was afterwards sent by Mr. Buck into Manbhum, and in Govindpur I spent a week amongst the Santals and Khols examining the possibilities of obtaining supplies of Tussur raw-silk and in inducing the natives to cultivate its production. I was the guest of the Rev. Mr. Campbell, a Scotch Missionary amongst these tribal and aboriginal people.

I afterwards went to Patna, Futwa, and Gya, and saw considerable quantities of Tussur cocoons being reeled by the natives, and collected examples of both cocoons and specimens of their reeling.

I took great interest in this sericicultural work. It caused me to consider whether there were no climates in India suitable for the cultivation of the European univoltine silkworm, the Bombyx mori, which is the worm from which is chiefly obtained the silk of commerce both in Europe, China and Japan; the silk of Bengal being the product of two or more species of multivoltine silkworms, Bombyx fortunatus, and Bombyx crœsi, species belonging to the more sub-tropical regions of India.

I was led to consider the suitability of Kashmir, where the climate in summer much resembles that of Italy, and therefore suitable for the cultivation of the mulberry and of the univoltine silkworm of commerce, the Bombyx mori, and it was accordingly to that important State that I turned my attention, with the results I am about to describe.

CHAPTER II.

HISTORY OF KASHMIR SERICICULTURE. CORRESPONDENCE. FIRST PURCHASE OF SILKWORM EGGS. FURTHER INCREASED PURCHASES. SUCCESSFUL BREEDING. STAFFORD HOUSE SILK EXHIBITION, 1894. NECESSITY FOR EDUCATIONAL TRAINING IN SERICICULTURE.

IT is a great pleasure for me to put on record the names of those with whom I have had correspondence and who have assisted in helping forward the phenomenal development of this now important and useful Industry.

I will do so in order of chronological priority.

In June 1889, I received a letter from my friend Mr. John Lockwood Kipling, then Director of the Art School at Lahore, whose guest I was in 1886, and who wrote to me that year about Kashmir Silk when he was visiting Colonel Parry Nisbet, C.I.E., the British Resident at Srinagar.

This prompted me to write to Colonel Nisbet, whom I had met in Peshawar, when he was Commissioner there, but who had since been transferred to Kashmir.

This was some time before I had any official correspondence on Kashmir Sericiculture with the

Government of India in India, or with the India Office in London.

I at once entered into correspondence with Colonel Nisbet, who corroborated Mr. Kipling's information, and looked favourably on my idea of the possibility of developing a useful industry in the production of raw-silk.

The following letters will be sufficient to show the initial stage of thought and action at that time. I cannot find Colonel Nisbet's letters to me; they were probably destroyed with my office letters, as is my custom every seven years. I have written to Colonel Nisbet to ask him if he retained copies of them, but he informs me he did not.

[COPY.]

Leek, 6th January, 1890.

Dear Colonel Nisbet,

Some time ago I received a very interesting letter from my friend Mr. Kipling, who wrote to me when your guest, from the Kashmir Residency.

He draws a very delightful picture of the capabilities of Kashmir for vegetable products, and especially mentions silk.

I am very glad indeed to hear it. It was one of my regrets on leaving India that I had not time to go to Kashmir, for I was extremely anxious to have acquainted myself with its capabilities as a silk-producing State, because I had heard a good deal of it in times past, and of several unsuccessful sericicultural efforts.

The result of my Report on sericiculture in Bengal to the Government of India caused a demand for Bengal silk, which in Leek soon quadrupled after I was there. The price has risen considerably, and much more than is grown could easily be sold, and I bring this before your notice in order to point out what a splendid opportunity there is in Kashmir, now there is a probability, under your influence and good management of its becoming in time a great silk-producing State.

I remember when I visited Peshawar in 1886, I found a good deal of Bokhara silk in the bazaars, and I wondered that Indian silk was not used.

I think it would be of great importance to have some examination made as to the silk-producing capabilities of Kashmir, and to establish there such a Sericicultural Institution as that at Padua, and the one at Montpellier, both Government Institutions, which are doing immense good to sericiculture in Italy and France.

It always seems a great pity to me that Europe, and especially England, does not get more Silk from our own dependencies. We have to go further afield, chiefly to China and Japan, for the bulk of what is needed.

I thought I would venture to trouble you with these observations in the interests of such a beautiful industry as silk, and which would benefit both England and Kashmir.

I feel convinced that a very large silk-growing

industry for export as well as for native use, could be established in Kashmir.

In 1891 we shall have an Exhibition of Silk in London, as you will see by the enclosed paper, and as President of the Silk Association of Great Britain and Ireland, I should be very happy to see Kashmir well represented as a silk-producing country.

I shall be very glad at any time to be of service to you in matters relating to Kashmir Silk.

Believe me,

Yours very truly,

(Signed) THOMAS WARDLE.

[COPY.]

Leek, June 24th, 1890.

Dear Colonel Nisbet,

Yesterday morning my wife handed me a letter dated the 16th November of last year, which I had not sooner seen, for during my absence from home letters were placed on my dressing-room drawers to wait my return, and this one was found on cleaning them at the back on the floor.

I am very sorry indeed it has escaped notice so long. Probably my not replying to this may have been the reason I have not received any reply to my letter of the 8th January.

It is quite correct as you state that the London Silk Exhibition of this year was to be only for British and Irish Silks. The object was to show ladies that Silks are and can be made in Great Britain, equal to their requirements, and it was only a tentative Exhibition to that end; but it is intended in 1892 to have a large Exhibition at the Imperial Institute, or elsewhere, in which India is to be fully represented. Then will be the time for you to show us what Kashmir can do.

Our Exhibition, the one held at the town house of Lady Egerton of Tatton, and opened by the Duchess of Teck, passed off very successfully indeed. A National Silk Association of Ladies has been formed, of which I send you a circular. I hope Mrs. Nisbet will sign it, and get other ladies to do so, and send it back to Lady Egerton of Tatton, the Hon. Secretary.

Already we are making Spitalfields busy.

I send you an account of the Exhibition which appeared in most of the papers.

I am deeply interested in the Silk future of Kashmir. I feel quite sure from its physical geography and beautiful climate that it is destined to be a great silk-producing country. It is almost the only country I know where the Bombyx mori is found wild in any numbers, and the wild cocoons collected. I think the climate is suitable for the development of the best races of Bombyx mori, such as those of Italy and France, and it ought to have a Government Sericicultural Laboratory, not a State one, but an Imperial one, and under Imperial management and control.

If I could do anything to assist the progress of Kashmir sericiculture I will very gladly come out and report on its capabilities etc., if you wish me to do so.

The Silk Association has just published its Report of last year and part of this. I have much pleasure in asking you to accept a copy. You will see in it a copy of my correspondence with Lord Cross.

With kind regards to Mrs. Nisbet,

Believe me,

Yours very truly,

(Signed) THOMAS WARDLE.

[COPY.]

Leek, October 27th, 1890.

Dear Colonel Nisbet,

I have just received the samples of Kashmir raw-silk with which I am delighted. I will have their value thoroughly assessed, and will fully report upon them in the course of a very short time.

Can you inform me whether they are the produce of wild Kashmir Bombyx mori cocoons, or of the cultivated or fed worms?

I am told that there are people in Kashmir who pay the Government an annual sum for the privilege of collecting from the mulberry tree the cocoons of the Bombyx mori.

I fear the prices at which the samples are marked will be prohibitive, being much dearer than Italian raw-silk in London.

Can you send me samples of other qualities of

raw-silk? I cán guarantee a considerable consumption of them in Leek from the manufacturers of sewing-silks, if their value could be about 10/- to 12/- per lb.

Yours very truly,

(Signed) THOMAS WARDLE.

[COPY.]

Leek, 5th December, 1890.

Dear Colonel Nisbet,

With further reference to your letter and its accompanying parcel of samples of Kashmir Silk, I have carefully examined them, and have also submitted them to two gentlemen of the greatest experience; one a London silk broker, and the other a silk manufacturer, and I send you their opinions, which I entirely endorse, and they leave me very little to say.

I strongly urge the adoption of an improvement in cocoon-reeling on European lines.

There is not the slightest doubt that it might be the foundation of a large and remunerative industry, and with such a beautiful climate as Kashmir is favoured, it seems to me that it would add to the immense usefulness of India to us and to herself if Kashmir could be made to supply us, and also France, with such beautiful silk.

I am just considering the order for a set of cocoon-reeling machinery consisting of a number of reeling bassines and equipment at the request of the Government of India, from France and Italy, and have received the estimate. As it may possibly be of interest to you to know the particulars, I have much pleasure in sending them.

Yours very truly,

(Signed) THOMAS WARDLE.

It may be well to mention here some notes with reference to the earlier history of Sericiculture in Kashmir.

In Sir Walter (then Mr.) Lawrence's interesting book "The Valley of Kashmir," written by him while Settlement Commissioner in Kashmir, the following information is given:—

"The history of sericiculture in Kashmir has been fitful and desultory. The silk industry is of ancient standing, reference being made to it in A. D 1536, by Mirza Haidar, in his History."

"Before 1869 the industry had existed in the unorganized, crude state in which it had probably existed for centuries, but nothing however is known in Kashmir about the origin of its silk industry, beyond the fact that it is very ancient, and that it is intimately connected with that of Bokhara, with which it has always had interchange of seed and silk."

"In later times the important date is the year 1869, when Maharaja Ranbir Singh, an enthusiast in new industries, revived the silk production on a large scale. The cost of buildings and plant was enormous, and the rearing-houses being scattered in all parts of the valley, could not be properly supervised. There was no one possessing any technical knowledge to supervise, and though great improvements were made in reeling, there was no man in Kashmir who could avert the calamity which befell the industry in 1878, when nearly the whole of the silkworms were carried off by disease. Every credit is due to Babu Nilamber Mukerji, the Chief Justice of Kashmir, for his efforts and his success in improving the reeling of the silk. The industry lingered on till 1882, and from that time to 1890, the State had left it to the silk rearers; the quantity of seed rapidly diminished, and sericiculture was virtually at an end."

Why did these efforts not succeed? That is easy to tell.

Ignorant procedure. Letting silkworm diseases have their own way, because they were not understood. Ignorance of the scientific work of Pasteur, and of the present methods of prevention of disease, which he alone first formulated.

The Lister Dehra Doon Grant failed in the same way, at a loss of £50,000. There need not have been a loss, but a big profit, had all the conditions of the problem been understood. Wherever the mulberry will properly grow, silk can be profitably produced.

When in Bengal in 1886 I found a mortality of 60 per cent. of the silkworms there through quite preventible disease. My recommendation of an Imperial sericicultural station passed unheeded, and I doubt whether much material improvement amongst the native silkworm-rearers has yet been effected.

In 1889 I also communicated with Mr. (now Sir E. C.) Buck, who at once took steps to further the idea of Kashmir Sericiculture, by sending over a Mr. R. Mukerji (a younger brother of Babu Nilamber Mukerji, mentioned by Sir Walter Lawrence in his book), to see what could be done. The operations in Kashmir from May, 1889 to 1897, when a new arrangement described below, came into force, were under his charge, in subordination to the Settlement Commissioner, but with very imperfect appliances if measured by modern scientific methods.

The possibility of a successful future for this Industry was first suggested to the Government of India by me on the 14th December, 1891, in a letter I wrote to Mr. (now Sir Arthur) Godley, Under Secretary of State for India.

Having for a number of years had the privilege of knowing Sir George Birdwood of the India Office, I also wrote to him urging the possibilities of Kashmir as a sericicultural country.

He at once took the matter up very warmly. I, indeed owe much to him for the frank, unreserved confidence he always placed in me and in my opinions, and several times I discussed the matter with him both by letter and orally; and very great credit is due to him for all he did in furthering my ideas, and actions. In short, from the first, and down to the time of his retirement from the India Office, he has spared no pains in giving me every encouragement in my work.

He directly conducted the whole of the India Office official correspondence with Kashmir from 1891 to 1902, on the subject.

In 1894 a few pounds of Kashmir raw-silk were sent over to England by Mr. Rishibar Mukerji.

In a letter I wrote to Sir George Birdwood on the 21st February, 1894, I said:—

"I propose if this eight pounds of Kashmir raw-silk comes in time, to have some of it manufactured into silk fabrics, and as the first fabrics made out of silk grown in Kashmir. I have a deep conviction that the future of the silk industry in Kashmir will be of great importance, and that if the Maharajah of Kashmir thinks favourably of the idea, Kashmir may soon become one of the greatest centres in the world for sericiculture. I ask for your earnest co-operation in this matter."

This silk was manufactured into brocade by Messrs. Warner & Sons, of Spitalfields and Braintree, and exhibited at the Stafford House Silk Exhibition in 1894. It was shown by me to the late Queen and to the then Prince and Princess of Wales, who evinced much interest in it.

It was described by me in the Catalogue of that Exhibition as follows:—

"R. Mukerji,
Director of Sericiculture, Srinagar, Kashmir,
Small experimental Silk cloths, manufactured in London from the first raw-silk from Kashmir."

Mr. Mukerji's connection with sericiculture terminated at the beginning of 1894, and Mr. W. R. (now Sir Roper) Lawrence was placed in charge. In July, 1895, some further samples of Kashmir raw-silk were brought over to London, on behalf of the Kashmir Durbar, to Sir George Birdwood, and by him forwarded to me for report as to quality, etc.

My report was sent to the Government of India by Sir George Birdwood, on the 5th December, 1895. In it I advised that the stock of raw-silk retained by the Durbar, pending the examination of the samples sent over, should be shipped direct to London and sold there; and on the 29th November, 1895, Sir George Birdwood was informed that, under the orders of the Durbar, sixteen cases containing sixteen bales of raw-silk had been sent, via Calcutta, addressed to him for sale in London. These cases were received on the 26th January, 1896, and samples were at once forwarded to me for distribution amongst silk manufacturers, merchants, etc. Their various reports on the raw-silk were fairly satisfactory, but recommending considerable improvement.

Here I must not omit to mention the name of my most excellent friend Sir M. M. Bhownaggree who for many years has been watching the industrial interests of India.

I well remember in an excellent speech at the annual dinner of the Silk Association of Great Britain and Ireland, at the Freemason's Tavern, London, in 1891, he used the following prophetic words:—

"There is in India any amount of raw material, for instance there is in Kashmir abundance of raw-silk, and if efforts like those of Mr. Wardle and this Association could teach her how to make use of it, India can supply not only her own wants, but those of other parts of the world."

These words sank deeply into my mind, and I think it will not be denied that they have borne good and abundant fruit.

I do not see why, with the climatic potentialities of a great country like India, we should go to the farther East for silk, or any other tropical materials, cotton for example, which with proper knowledge, could no doubt be as well, if not better, grown in India than anywhere else.

In 1896 Sir Adelbert Talbot became Resident in Kashmir, and, as a result of his study of my encouraging Reports on the possibilities of Kashmir silk, he recommended His Highness the Maharajah of Kashmir and Jammu to recommence the Industry which had become extinct, on a commercial scale and in a more scientific and extensive manner.

In Chapter 8 I have referred to the active part taken by Sir Adelbert Talbot, in his correspondence with Sir George Birdwood on the point, and to the lengthy correspondence which ensued between the India Office and myself, recorded in the memorandum of the engagement of a Director of Sericiculture, and the purchase of silk machinery, for the Kashmir State, in 1896-7.

The Kashmir State Council accepted Sir Adelbert Talbot's advice and the Government of India were addressed accordingly.

*Captain J. L. Kaye who had succeeded Mr. Lawrence in 1895, was in 1897 the Settlement Commissioner of Kashmir. He at once took a great interest in sericiculture, an interest which has been continued during his stay there down to his retirement in the early part of the present year.

Sir Walter Lawrence in concluding a very interesting account on pages 368 and 369 of his charming book "The Valley of Kashmir," wrote :—

"It is impossible to exaggerate the potentialities of silk in Kashmir," but he suggested it should be left to individual enterprise to develop it and not be a State-guided industry.

With his first sentence I am in thorough agreement, but not in his opinion that the industry should not remain in the hands of the State.

Since he wrote his book the progress made in the industry has verified my view, that it ought to be managed by the State, but things were not so hopeful when Sir Walter wrote in 1895 as time has since proved them to be : in fact everything was purely experimental.

It was through Sir George Birdwood that I received instructions in 1897 from the Secretary of State for India, to go to France and Italy accom-

* Captain (now Major) S. H. Godfrey acted for Captain (now Major) Kaye as Settlement Commissioner of Kashmir for a year in 1898-1899.

panied by Captain Chenevix-Trench, then the Assistant Resident in Kashmir, for the purpose of purchasing silk-worm eggs and cocoon-reeling machinery.

On my return from the continent in June, 1897, I sent in my report of what I had seen and done, of the cocoon-reeling machinery and the silk-worm eggs I had purchased, the latter amounting to over £600 worth. This Report will be found in a following Chapter.

On my return, Sir George Birdwood becoming acquainted with the extent of my purchase of silk-worm eggs, from various silk-worm breeders in Italy and France, said he feared I had purchased more than the Government of India would care to be responsible for. I offered at once to take the responsibility, for I knew from experience, and in the care I had exercised in purchasing eggs free from disease, that the time was ripe for practical work, and that that for experiment had passed.

I was much gratified in due time in learning from Kashmir that the hatching of these eggs and the rearing of the worms had been completely successful and free from disease, and I was requested to purchase £1,500 worth for the ensuing season. I did so and success fortunately also attended that purchase. The following year (1899) £3,040 worth were ordered.

The actual produce of cocoons from this purchase was 12,000 maunds or 984,000 lbs. and of raw-silk 1,000 maunds, or 82,000 lbs.

One pound of cocoons yields one ounce six drams of raw-silk, which is about the normal production in South Europe.

The success in the rearing of silk-worms from this latter purchase of eggs proved so great that orders arrived in 1900 for the purchase of 25,000 ounces. This quantity cost £4,000, and the out-turn from these was 1,053 maunds of raw-silk or 65,346 lbs., but with greater care and experience in rearing of the worms, and good weather, upwards of 25 per cent. greater yield ought to have been realized.

Mr. Walton gave the 25,000 ounces of seed purchased in 1901 to 6,000 householders. They farmed it out, and in fact more than 25,000 people were employed in rearing the crop throughout the valley. Thus the production of the silk is essentially a cottage industry, as far as silk-worm rearing goes; and the reeling of the cocoons is entirely done by Srinagar workmen and boys in the new filatures.

During my stay in Kashmir I advised that the purchase of eggs might safely be increased by 5,000 ounces, and it was at once decided to purchase 31,000 ounces for an increased crop of silk for the present year (1904) and in addition 1,000 ounces of eggs of the value of about £133, from Italy, for trial to obtain good commercial white raw-silk like that of Japan.

I have recommended the Government of India to adhere steadfastly to the purchasing of undiseased eggs from Europe, notwithstanding advice to the contrary by those who erroneously thought that "eggs of reproduction" in Kashmir would do as well. Such advisers did not seem to think that silk-worm breeding and rearing is in Europe a very scientific industry, and that it is impossible to expect the village Kashmiris to understand the use of the microscope in detecting pebrinised moths and eggs, or in the various minutiæ of this branch of the industry, which is so completely understood, and so skilfully practised, in Europe.

But nearly every year since 1897, at my recommendation, such eggs of native reproduction have been tried experimentally in Kashmir on a small scale, and have through want of knowledge, always failed. From a Report of the Proceedings of the fourth Conference on Sericiculture, held at the Srinagar Residency, on the 22nd October last year, which I have received from Mr. Colvin, the present Resident, I am informed that it has been decided that the danger of using tainted seed could not be over-estimated, and that the cocoons raised in the Ramban Tahsil in 1903 as a further experiment, had been subjected to microscopic examination, and the seed again found to be tainted with disease and unfit for distribution as industrial seed. Directions had accordingly been given that these cocoons should be used for reeling only, and not for seed.

It will be seen from all this how completely I was justified in making what was thought at the time my over-adventurous purchases of seed in 1897 and the following years.

I therefore most earnestly recommend a continuation of the system hitherto adopted, and which has proved so wonderfully successful; and that no official nervousness may ever be allowed to mar the results secured by my decisive action in 1897.

I venture to add, see Chapter 7, a copy of the correspondence between the Government of India and myself in 1889; a correspondence initiated by me with a view of securing for India under Imperial control (rather than under Presidential), an educational establishment of sericiculture both for India and Kashmir. It is to be regretted that this advice was not accepted, but that instead a laboratory under the direction of a mixed committee in the Bengal Presidency was adopted.

My idea was to see established an Imperial Government Institution with branches such as those that have been in operation for a long time in France and especially in Italy, of which an account will be found in Chapter 17; where every department of scientific sericiculture is conducted, and taught by highly qualified technical savants, resulting in annually improved breeds and races of silkworms, improvements in the cultivation of the mulberry, in the physiological study of the origin and prevention of the diseases peculiar to silkworms—pebrine, flacherie, muscardine, grasserie, surcina—as well as of all matters pertaining to scientific and commercial Sericiculture, and in the thorough technical training of young men intending to enter into this important branch of Agricultural Industry.

I have no hesitation in stating that if my suggestion had been acted upon, not only would sericiculture in Bengal have been vastly improved, but would have by this time become an important industry in other parts of India, such as Hyderabad, Mysore, etc., where the mulberry can be well cultivated, and where there has always been some silk-production and silk-weaving done.

As it is, however, Kashmir is happily the pioneer, and the marvellous success of the efforts there adds weight to my previous advice, as well as shows that we need not be so dependent on other countries for supply of raw material, whether silk—or other fibres—if we will look scientifically after the resources of our own possessions.

CHAPTER III.

ENGAGEMENT OF A DIRECTOR OF SERICICULTURE FOR KASHMIR. BUILDING OF FILATURES. PRODUCTION OF RAW SILK, 1898 TO 1904.

When the Secretary of State for India sent me to France and Italy in 1897 to make enquiries about sericiculture, and to purchase silkworm-eggs and cocoon-reeling machinery for Kashmir, I was also requested to engage the services of either a Frenchman or an Italian, who would occupy the position Mr. Walton has now so worthily filled during the last few years.

I made enquiries, but was not satisfied, and I returned to London to report that I preferred to decline the responsibility of engaging anyone, either French or Italian, because none of them understood either English or Hindustani, and besides I feared such an appointment, if made, might prove unsatisfactory.

Happily I found Mr. C. B. Walton (son of the late Lieut.-General Walton, C.I.E., S.C.) whose acquaintance I made at Surdah during my first visit to India in 1885-6, and who for twenty years occupied important positions at the Surdah filature in Bengal, six years as Assistant, eight years as Manager, and six years as Director of the three concerns of Surdah, Maldah, and Moorshidabad, and was familiar with the language, was disengaged. I unhesitatingly recommended him for the appointment.

He was at the time open to engagement, having retired from the Directorship of the Bengal Filatures, and was living with his wife and family at Musoorie.

The Under Secretary of State for India requested me to telegraph to him at once to go up to Srinagar to see Sir Adelbert Talbot, the then Resident, on whose advice the Kashmir Durbar gave him the appointment on the 1st May, 1897. He still holds it, and the selection has been a very happy one.

The erection of filatures was at once commenced and was continued from year to year, six having been completed by the close of Sir Adelbert Talbot's tenure of office, towards the end of 1900, and four new ones commenced in 1902 being almost ready for use when I left Srinagar in May, 1903. Each filature is 425 feet long, and contains 212 bassines, or in all 1272. The building of the first six filatures cost £28,000, and in all this work Mr. Walton was warmly supported and directed by Sir Adelbert Talbot, who continued to give personally great attention to this effort.

When Mr. Walton was appointed in 1897, the out-turn of raw-silk was for the year preceding only 10 to 12 maunds (820 lbs. to 984 lbs.)

In 1898 he had the first European seed (purchased

by me in Italy and France in 1897) to work upon, but there was heavy expenditure in this year, and the 1½ maunds of seed (1,968 ounces) which cost £630 did not make a profit, as Mr. Walton kept some 200 maunds of cocoons (16,400 lbs.—a Kashmir maund being 82 lbs.) for trial, on a practical scale, of "reproduction" for the following year.

If he had not done this there would have been no loss.

There was a considerable loss in the exploitation of this seed for "reproduction," in fact as much as 50 per cent. of the eggs were unproductive, chiefly through the ignorance and inexperience of the natives in sericiculture, and partly through climatic conditions being unfavourable.

Another cause has been that the mulberry trees had been allowed to grow without careful cultivation, and had never been pruned. The consequence was that the leaves were not in the best state for feeding the worms.

Unfortunately these cocoons of native reproduction proving, as I fully expected, highly pebrinised, failed to bring forth healthy moths, or productive eggs.

The year 1899 showed a handsome profit on 5 maunds (6,560 ozs.) of silkworm-eggs; in 1900 the profit was more than in 1899, but did not offer a fair comparison with that year, as the market in 1900 was low and the silk sold very badly, in fact only realising from 11s. to 12s. 6d. per lb.

The raw-silk which was sent to England in 1899-1900 was about 150 bales, or 22,500 lbs., which fortunately found purchasers up to 16s. 9d. per lb.

The total value of this would be about £18,844, and in addition there was the value of the waste-silk or "déchets" resulting from the reeling of the cocoons, the current price for good waste-silk in 1900 was 3s. per lb. and the proportion of waste being about one-third, made an additional total of about £2,145.

Owing to the more trained manipulative skill on the part of the cocoon-reelers, the raw-silk is beginning to and should soon approximate, if not equal, that of Italy and France, and consequently command a corresponding price. In 1902, Mr. Walton sent over to England 953 maunds (74,186 lbs.) or 522 bales of silk, (a Kashmir bale weighs 145 lbs.), besides sending a 100 maunds (8,200 lbs. or 54 bales) to the local markets for local use. The waste, which as I have already stated is generally one-third of the quantity reeled, is of two qualities, and realised respectively 2s. 8d. and 1s. 7d. per lb. There is also the silk from the inferior and double cocoons, which are sorted out; this amounted in 1901 to 100 maunds (82,000 lbs.) and fetched from 25,000 to 30,000 rupees (£1,666 to £2,000).

This quantity of silk (1,053 maunds, or 86,346 lbs.) is the product of 20 maunds (1,640 lbs.) of eggs. One ounce of silkworm-eggs weighs 30 grammes, and contains from 30,000 to 40,000 eggs, and yields 42 kilos or 92½ lbs. of fresh cocoons. A French ounce is 482·5 grains.

My sole reasons for recommending these extensive purchases of European seed were on account of the difficulty of obviating disease in Kashmir-raised seed, and of the security by purchasing immune eggs, besides the undoubted superiority of eggs bred from the best races of silk-worm moths from European localities where silk-worm breeding is conducted as a science.

The present growth of the black mulberry tree (Morus niger) in Kashmir available for silkworm culture is calculated to be sufficient to produce annually 50 lakhs worth (£333,333) of silk. But I have strongly recommended the systematic planting and cultivation annually of the white mulberry (Morus alba) such as is grown and preferred in France and Italy, when, if adopted, the area will be extended, the production of raw-silk greatly increased, and a valuable and important industry consolidated in the beautiful climate of Kashmir.

Reckoning the rupee at 1s. 4d. which is the present rate of exchange, and likely to remain so, the amount would, as stated above be £333,333 worth per annum. So it may be safely assumed that the outlook for the production of raw-silk in Kashmir is of a very promising nature.

The following table compiled during my stay in Kashmir, gives the number of Silkworm Rearers employed in the villages of the Valley of Kashmir, and the quantity and value of seed used yearly since 1897, as well as the quantities and values of the yield of silk per year, and the number of persons employed in rearing, not including the factory or filature cocoon reelers :—

Year.	Quantity of Eggs in ounces imported.	* Ditto Local.	Value of Eggs and Carriage.	Raw Silk produced.	Waste Silk produced.	Local Silk & Waste Inferior Cocoons.	Total Value of Out-turn.	Number of Rearers who took eggs.	Average Number of Rearers engaged.	Quantity of Fresh or Green Cocoons.
			£	lbs.	lbs.	lbs.	£			lbs.
1898	1,920	...	766	5,412	1,863	351	4,858	326	1,304	1,07,100
...	...	6,400	...	Nil	Nil	Nil	...	1,106	4,424	Nil
1899	6,400	...	1,632	22,509	7,500	7,239	...	803	3,212	3,26,672
...	...	12,800	...	560	320	500	14,684	2,255	9,020	16,000
1900	19,060	Nil	3,000	44,181	23,063	7,739	27,419	4,290	17,160	8,13,792
1901	25,606	do.	4,333	65,931	32,870	12,346	49,118	5,887	23,548	10,65,204
				Estimated at						
1902	25,527	do.	4,333	†1,20,750	†54,500	†12,000	†96,666	†8,158	32,632	†18,82,692
1903	25,521	do.	4,333	‡ ...	‡ ...	‡ ...	‡ ...	11,060	44,245	...
1904	35,500	do.	5,500	...	...	...	...	...	...	...

* The Eggs were of local reproduction, both moths and worms died of Flacherie and Pebrine.
† The exact figures to be obtained later on.
‡ Not yet realized but anticipated to be very satisfactory.

The eggs in each year were of course purchased the year previous to the production of the raw-silk, so that eggs put down for the present year were really purchased last year, *i.e.* for the silk crop of 1904.

The above mentioned rearers are householders whose families assist in silkworm-rearing. Mr. Walton thinks that the total number of persons, young and old, averages 4 to 5 persons in each household, thus then giving employment to from 50,000 to 60,000 people. One ounce of eggs is given out to each family of 4 or 5 persons, and two ounces to larger families.

The value of the silk, with the waste, is shown above to be from £90,000 to £100,000, a most valuable addition to the revenues of the Kashmir State, and to the material prosperity of the villagers who are liberally paid for silkworm rearing by receiving a very remunerative price for the cocoons they bring in to Srinagar. The eggs are distributed to them gratuitously.

The 20 maunds (30,000 ozs.) of eggs of the value of £4,000 purchased in 1902 were distributed over a larger number of men, most of whom have now gained experience, and, the trees being pruned each year, will yield much better results as regards the quality of the cocoons, as well as the greater out-put of raw-silk.

My correspondence with Mr. Walton has been very voluminous. I have spared neither time, trouble, nor expense in endeavouring to obtain every possible kind of information about European sericiculture, as well as devoting myself to persuade silk manufacturers to forego their prejudices and give Kashmir silk a fair and impartial trial.

In May last I went to see Messrs. Chabrières, Morel et Cie, of Lyons, the largest distributors of Raw-silk on the Continent. They then held 60 bales of Kashmir Silk; they reported of it most favourably, and told me it was fast becoming sought after by some of the principal manufacturers of France, and that they could dispose of any quantity Kashmir could produce for some time to come.

Another silk merchant of the firm of E. Lacharriere & Cie, 4, Rue Desirée, Lyons, came from Lyons to see me last October, earnestly wishing to do business with Kashmir direct, and reported in the same strain as Chabrières, Morel et Cie.

I had to inform him that the Durbar had decided to employ only Messrs. Durant, Bevan and Co., 9, New Broad Street, London, E.C., the well-known and oldest established silk brokers in Europe, to distribute the silk at the reasonable brokerage of 1 per cent., and I referred him to them.

In England also I am receiving very favourable reports, and I feel sure that as soon as the cocoons can be produced and reeled as perfectly as they are in France and Italy, the silk will be of the same high value. Up to the present time it realises prices from 1s. to 2s. per lb. less than the best Italian silks, so far a very satisfactory result, considering the number of persons occupied in reeling, in 1904 over 5,500, who were comparatively new and inexperienced in the extremely delicate manipulation of cocoon-reeling.

I feel sure that were it not from the innate delicacy of touch and their skilful fingers, these men and boys would not have become so proficient in as many years as it has taken many of them months to learn to reel.

Mr. Walton obtained in 1902 one maund (his cocoon maund is 88 lbs.) of cocoons per ounce of eggs against half maund (44 lbs.) the previous year, and last year he hoped to get 110 to 132 lbs. from an ounce, as the reelers gain more experience.

The last Resident, Mr. Louis W. Dane, and the present Resident Mr. E. G. Colvin, have followed up Sir Adelbert Talbot's successful efforts with great ability and energy, and the outlook now under Mr. Colvin's able guidance is exceedingly encouraging.

I have much pleasure in reporting that the successful development of sericiculture in Kashmir in all its technical details and management, is greatly due to Mr. Walton, the Director, for the perseverance, knowledge, and skill with which he has with untiring energy and firmness conducted this industry during the seven years of its existence. I was greatly pleased and satisfied with the efficiency with which the six large filatures at Srinagar were being conducted, and with the enormous number of native people employed in them reeling cocoons. The new Industry has been fortunate in having Mr. Walton for its Director,

This Industry has also the advantage of a Conference Committee, which meets periodically. It consists of General Rajah Sir Amar Singh, K.C.S.I.; the Revenue Member of the State Council; the Governor of Kashmir; the State Medical Officer; the Chief Judge; the State Engineer; the Accountant-General; the Settlement Commissioner; the Director of Sericiculture; the Assistant Resident; and the President, Mr. Colvin, Resident.

CHAPTER IV.

SILK SECTION OF THE EARL'S COURT EXHIBITION, 1900. CONSIGNMENT OF SILK, BOMBAY VERSUS CALCUTTA.

IN 1900 I arranged a second Exhibition of Kashmir raw and woven Silks in the Silk Section of the Woman's Exhibition, at Earl's Court, which was described by me as follows:—

KASHMIR SILK.

Exhibited by His Highness the Maharajah Sahib of Jammu and Kashmir, India.

1. Raw-silk, Organzine, and Tram "in gum" undyed.
2. Figured and plain Brocades, Brocatelles, Damasks, etc., all manufactured from the raw-silk produced in Kashmir, in 1898 and 1899, for Upholstery, Decorative and Dress purposes, in English, Indian, Italian, Venetian and French styles and designs. The Silk dyed with pure unweighted dyes by Messrs. Joshua Wardle and Sons, Silk Dyers, Leek, Staffordshire, and woven by Messrs. Warner and Sons, of Braintree, and Spitalfields. This most interesting Case of Silk of Kashmir

production is illustrative of the successful attempt within the last few years to introduce Sericiculture into Kashmir, the climate of which is extremely favourable both for the rearing of the silkworm, and the cultivation of the mulberry leaf, upon which it feeds.

The illustration of the Case of Kashmir Silks exhibited was made for the Earl's Court Catalogue.

When I at first recommended the Government of India to take up sericiculture in India, I advised that the raw-silk produced there should be sent direct to London, via Karachi or Bombay in preference to Calcutta. My chief reason for this was that knowing that Bengal silk was not used for the great bulk of European textile necessities, China and Japan silk being chiefly required, I felt convinced that prejudice might probably arise in the minds of European silk manufacturers if any idea of Bengal silk should be entertained.

It was thought, however, that it might be better to consign the silk to London by way of Calcutta, that hitherto having been the usual channel for the silk exports of India.

But here I would like to observe that Bengal silk has very important uses, and I would not like to be understood as in any way disparaging it. It is preferable to that of the silk of other countries for certain purposes, but these are limited in comparison with the uses of the silks of China, Japan, Italy and France.

PLATE I.

Kashmir Silks at the Earl's Court Exhibition of 1900.

The high reputation of Bengal silk filatures is of historic importance, they have long been celebrated, and will doubtless have as at present a continuing and increasing future.

In the Appendix I have given a list of the Bengal Filatures as they existed in 1901.

With the continued and increased application of the aids of modern sericicultural science, there is no reason why Bengal silk should not acquire an ever increasing reputation and expansion as well as also in Mysore, Madras, and other parts of India, where the climate is suitable to the cultivation of the mulberry whether grown as a tree or a shrub.

I found the prejudices I had anticipated did arise, and with a view of meeting them I visited at considerable expense, trouble, and anxiety, most of the silk textile users in this Country, as well as several large Continental manufacturers, and succeeded in obtaining for the new silk of Kashmir a more favourable consideration, but even now much prejudice exists.

When samples were first put before them they said, "Oh, we know Indian Silk very well; it will not suit our textiles; we have repeatedly tried it." They had conceived a strong prejudice against it, thinking it must be of the same class and character as that of the multivoltine sub-tropical silk of Bengal. The prejudice aroused nearly spoilt its chances; it certainly delayed them for some time, and even now it is far from being overcome by some manufacturers. This prejudice was not confined to England; it was shared in on the Continent in the same way, and it cost me much time and money to go over to Germany and France with samples to convince the manufacturers and dealers that it was not Indian silk at all, but that of the very best race of silkworms of France, but grown, so to speak, in Kashmir from eggs imported from the South of France, as well as some from Italy.

In time, however, the silk has been tried on its merits, and is now in demand. It is worth, as I have previously stated, from 1s. to 2s. per lb. less than standard French and Italian raws, but there is no reason why its quality and value should not equal the European silks as soon as cocoon-reeling in Kashmir, and other points suggested in my No. 2 Report, Chapter 15, have become more perfected, and this is rapidly being done.

There were also important reasons, geographical and others, for Kashmir silk being imported direct. This for the last two years has been done with marked beneficial results. It is now consigned direct by the Kashmir State to Messrs. Durant, Bevan & Co., who have taken the greatest interest in distributing it, chiefly in France, and also in England. Mr. Francis Durant, the experienced head of the firm, accompanied me to Srinagar, where he stayed a short time and fully explained what was required in the European silk markets to bring up Kashmir silk to the required standards of quality.

His visit had a very useful effect for no one in Europe better understands raw-silk.

The following Tables were furnished to me in Srinagar :—

	Rs.
Saving in commission and expense in 1901	20,000
Ditto ditto ditto in 1902	15,000
Prospective saving in the rest of the unreeled crop	25,000
	Rs. 60,000

Rps. 60,000 are equal to £4,000.

Table, in English money, showing the amount of net profit to the State in consequence of the change and of the advance in the price of silk.

	£
Saving in commission and expenses ...	4,000
Ditto, in increased price of raw-silk of 1901, 65,931 lbs.	7,897
Ditto in 1902 on 56,662 lbs.	8,735
Prospective saving on the unreeled crop of 1902 if the higher prices continue, estimated at...	10,000
	£30,632

CHAPTER V.

INDIA OFFICE CONFERENCE. STATE VERSUS PRIVATE ENTERPRISE. MY VISIT TO KASHMIR IN 1903. JAMMU DURBAR. SRINAGAR SILK CONFERENCE. SUCCESSFUL THREE YEARS' BALANCE SHEET AND PROFIT.

ON the 26th September, 1901, I was invited by Sir George Birdwood, K.C.I.E., M.D., LL.D., C.S.I., Special Assistant in the Revenue and Statistics Department of the India Office, to an informal Meeting at the India Office, which he had arranged in order to give me, Sir Adelbert Talbot, K.C.I.E., late Resident in Kashmir; Mr. T. W. Holderness, C.S.I., I.C.S., Secretary, Revenue Department of the India Office; Mr. L. W. Dane, C.I.E., the newly appointed Resident for Kashmir; and two other gentlemen of great knowledge and experience in silk and Indian sericiculture, the opportunity of expressing our views on the present position of the Kashmir silk industry and on its future, chiefly for the information of Mr. Dane, who was proceeding shortly to India to take up the post of Resident in Kashmir, previously held by Sir Adelbert Talbot, and others.

I was appointed Chairman, and a long discussion ensued as to the best means of further developing sericiculture in Kashmir. A detailed Report of this Conference will be found in Chapter 11.

My methods and advice were criticised, and the Government was strongly advised to proceed on the lines of "eggs of reproduction."

I, as strongly, protested, and begged the Government to let the system I had adopted and carried through so suçcessfully for four years, be continued; in fact, as I said, "let well alone, and continue each year to purchase eggs from Europe of the best developed races of silkworms;" and this, chiefly because I had each year a guaranteed undertaking from M. Arbousset, of Alais, a celebrated sericiculturalist, and other graineurs in Italy, that the eggs supplied to Kashmir should be carefully selected from moths microscopically examined, and their eggs wholly free from pebrine.

I am happy to say the Government has continued to take my advice, with results the most encouraging, as the foregoing table shows.

There need be nothing strange or abnormal in sending for silkworm-eggs from such a distance when, from a report of Mr. Adams, Secretary to H.M.'s Legation in Japan, even as far back as 1870, it will be seen that Japan exported to Europe no less than 3,776 cases of cards containing silkworm-eggs for France, and 2,583 cases to Italy, or a total of 6,359 cases containing 1,390,500 cards, of a gross weight of 319,829 lbs., or an average number of 220 cards per case. In 1868 the quantity exported amounted to no less than

2,300,000 lbs. This shows that France and Italy replenish their races by regular imports of eggs from the East, and this is without reckoning eggs from China, the quantity of which is generally very considerable.

PRIVATE ENTERPRISE.

At the informal Conference to which I have just referred, we discussed the question of private versus State enterprise. I am and have all along been decidedly adverse to this industry being taken over as a private enterprise, or by any Company promoting schemes.

I am distinctly of the opinion of the present Viceroy, Lord Curzon, emphatically expressed in a Speech by him at Jaipur in April of 1902, in which he said :—

"There is no spectacle which finds less favour in my eyes, or which I have done more to discourage, than that of a cluster of Europeans settling down on a Native State and sucking from it the moisture which ought to give sustenance to its own people."

From the first, and before this emphatic pronouncement of the Viceroy, my humble efforts for Kashmir have been wholly in this direction, and I have fearlessly opposed schemes projected both in London and in Srinagar, for private enterprise, being fully persuaded that the State of Kashmir, having had the courage, mainly through the wise advice and perseverance of Sir Adelbert Talbot when Resident, to induce the Maharajah to make a commencement on a practical scale with the eggs I purchased in 1897, and having embarked in the construction of a large filature and out-works, and in building new ones each year at great State cost as the Industry increased, that any profit made should go into the coffers of the State; although I would here like to remark that the question of the employment of thousands of native Káshmiris, greatly needing occupation, and then being on the verge of starvation, had more weight with me than the profit-making. That this has been accomplished is very gratifying, and not only so, but as I have shown, a very large money profit, amounting to upwards of £40,000, has been made during the last three years, which profit has properly gone to the State. See balance-sheet on pages 52 and 53.

In this view I was supported first by Sir Adelbert Talbot, and during the last three years by Mr. Dane and Mr. Colvin. The Maharajah and his brother, General Rajah Sir Amar Singh, both strongly declaimed in the Durbar assembled to meet me at Jammu in the spring of 1903, against the handing over of the industry to private enterprise, when I purposely asked them for their opinion.

Fortunately there were two, if not three, opposing schemes more or less matured for private enterprise; they materially helped to neutralize each other, but the Srinagar attempt very nearly succeeded, because it was thought, and I believe stated, by the Accountant-General in office before Mr. Anderson, that the industry was a losing concern, and naturally the then Resident said it had better go into private enterprise.

A lease was drawn up and partly signed; it required only the signature of the Viceroy.

On learning this, being totally opposed as I have before stated to private enterprise, or any Company promoters, I at once wrote to the Viceroy, Lord Curzon, through his private Secretary, Mr. (now Sir) W. R. Lawrence, and he laid my letter before Lord Curzon, who refused to sign the lease, and it therefore became firmly rooted as a State Industry.

I hope it may permanently so continue.

Having had correspondence with Sir Adelbert Talbot during his term of Residency, on the Industry and its prospects, I knew from him that up to the time of his leaving Office at the end of 1900 the Industry was profitable. He had informed me that the estimated profit for the year 1899-1900 was £13,333. I was unable to account for the statement made later on that a loss had been actually incurred, and necessarily felt much alarm on the ground of my own moral responsibility in having advised the Government of India to commence this Industry.

My fears were ill-founded, as I shall proceed to show.

I decided to go out to Kashmir at my own expense to see into matters for myself, both on this account and the interests of sericulture

PLATE 2.

Major-General His Highness Sir Pratap Singh, G.C.S.I., the Maharaja Sahib of Jammu and Kashmir.

PLATE 3.

Palace of H.H. the Maharaja of Jammu and Kashmir at Jammu.

generally, and to strongly advise the desirability of teaching the natives to weave the silk of their own production.

The Government of India approved of my going out, and I was most warmly welcomed on my arrival at Jammu, both by His Highness the Maharajah of Jammu and Kashmir, and his brother General Rajah Sir Amar Singh.

I was instructed by Mr. Colvin, who had lately succeeded Mr. Dane as Resident of Kashmir, and whom I met on my arrival at Sialkot, and had a day's discussion with him, to call first at Jammu before I went up into Kashmir, in order that I might have an interview with the Maharajah, who was then staying at his Palace at Jammu.

A Durbar was held in the Palace at Jammu on the 13th March, 1903, to welcome me, and to hear what I had to say on the various subjects connected with the silk industry in Kashmir, and on the mode of distribution of raw-silk in Europe.

I had a most hearty reception, and there were present, in addition to the Maharajah and his brother, Sir Amar Singh, Khan Bahadur Ghulam Ahmad Khan, Revenue Member of the Jammu and Kashmir State Council; Rai Sahib Bhagat Narain Dass, M.A., Judicial Member of the Jammu and Kashmir State Council; Rai Sahib Dewan Daya Kishan Kaul, B.A., Private Secretary to His Highness the Maharajah of Jammu and Kashmir State.

Mr. Colvin, the Resident, having gone on tour in the Himalayas, could not be present, but I met him soon after at Srinagar on my arrival there.

The Maharajah commenced the interview by hoping I had had a pleasant journey, and saying he was exceedingly glad I had come all the way from England in the interests of Kashmir, and that he was most grateful to me.

After a little conversation with him I asked him if he would like me to explain the object of my visit. He replied he would be very glad.

After making a few remarks about myself saying that so far as I was concerned my visit was absolutely a disinterested one, and that to me no pecuniary advantage would accrue, I read to him the index of my explanatory notes of my week's stay in the reeling districts of the south of France, written during my voyage to India, reading also portions of the text where it was felt desirable to do so. The meeting lasted nearly two hours, everyone seemed most interested. The particulars will be found in my No. 1 Report (see Chapter 14) which, after having been listened to with much attention and interestingly discussed, the Maharajah said he considered it to be of great importance, and ordered it to be printed at once by the State press at Jammu. This was acted upon, and I corrected proofs within a few days after.

I showed them the French tavelettes, as used in the Cevennes, the raw-silk samples I brought from Lyons of French cocoon reeling, and the magnificent black figured brocades I had had woven

PLATE 4.

General Raja Sir Amar Singh, K.C.S.I., Brother of the Maharaja of Kashmir.

in England by Messrs. Warner & Sons, of Kashmir Silk. The Maharajah with great pleasure accepted my suggestion that he should present a piece to the Queen and another to the Princess of Wales and asked me to be the bearer of them.

The Brocades were forwarded by Mr. Colvin to the India Office in London, and on respective days appointed subsequently by the Queen and the Princess of Wales I had the honour of offering them for acceptance on behalf of the Maharajah.

It was impossible not to be struck with the genuine satisfaction and pleasure evinced both by the Queen and the Princess of Wales on the reception of these beautiful silks.

I introduced the subject of private enterprise to the Maharajah, and asked him whether it was at all his wish that the industry should be given over to private enterprise, or to company promoting, and whether he did not think it was much more desirable that it should remain a State industry.

He rose, and in the most emphatic manner declared that he had no wish whatever that it should go into private enterprise. He was very much opposed to it and said it was intended as far as possible to have it retained and expanded as a State industry.

Sir Amar Singh warmly endorsed this expression, and said there was no intention whatever, nor had there ever been, to hand it over to those who had been trying to lay hold of it as a private enterprise.

I left Jammu on the 15th March, and arrived in Srinagar on the 19th March.

I spent a few days in the filatures there with Mr. Walton, the Director of Sericiculture, inspecting all the reeling and other operations, and waited at the Residency for the return of Mr. Colvin, who was then away in camp.

On his arrival I discussed with him the proceedings of the Durbar and my stay in Jammu.

He called a Conference on Sericiculture at the Residency on the 1st May, 1903; there were present besides himself:—

Pundit Man Mohan Nath Kaul Sahib, the Governor of Kashmir; Major J. L. Kaye, Settlement Commissioner; Mr. F. Anderson, Accountant-General; Mr. Walton, Director of Sericiculture, and I attended by special invitation.

Particulars of this interesting Conference will be found in Chapter 12, and also a report of a a subsequent Conference, Chapter 13, which was held last October.

At this conference I strongly advocated the commencement of silk weaving, and related the discussion that took place at the Durbar at Jammu on the subject, both as to weaving and as to its being a cottage industry and as to the desirability of giving employment to women. I also advised the appointment of a Forest Officer to superintend the cultivation of the mulberry trees which I found in a very badly managed state.

I warmly advocated the desirability of having

a thorough financial investigation of the state of the business up to that time, April, 1903. The Accountant-General, Mr. Anderson, was called into Council, and it was decided that such investigation and a stock-taking should be at once entered into. I should say that no such investigation had taken place for the previous three years.

Mr. Anderson, in the course of the following fortnight, had in a most able manner gone into every minutiæ, and produced his balance-sheet, and the result was most satisfactory; the conclusions erroneously arrived at, to which I have referred, were disproved, and a magnificent profit of £40,240 8s. od. for the past three years working was put on record.

This balance-sheet was however a provisional one, because the quantity and value of silk then in the state of cocoons, could not accurately be arrived at, except as cocoons, and a valuation of an *ad interim* nature was taken, giving credit for value of cocoons as cocoons only, notwithstanding I urged that a higher value should be taken for them approximating to their yield as raw-silk. However, Mr. Anderson being anxious to be on the safe side, was very properly inexorable, and the much lower value as cocoons, was taken.

Now when it is considered that the yield of cocoons for 1902 had been so prolific, that only one-third of them could be reeled in the reeling season, before the winter set in, for want of room, and the filatures necessarily became closed, we may safely assume that the profit has becom greatly increased.

The present value (February, 1904) of cocoons being 4s. 6d. per lb., and that of the raw-silks of Italy and France averaging 19s. 6d. per lb., gives a difference of stock value, less the expenses of cocoon-reeling, of a large amount, and we may therefore safely assess the item of profit at vastly more than £40,240 8s. 0d.; and in giving below a copy of Mr. Anderson's provisional balance-sheet for the financial year ending 12th April, 1903, I add a statement from a letter I received from Mr. Colvin, dated the 7th January, 1904, to the effect, that, although the next balance-sheet will not be out until the close of the financial year, April, 1904, it will show an enormously increased profit, although the drop in the silk market severely affects the result.

PRELIMINARY BALANCE-SHEET OF THE SILK INDUSTRY FROM SAMVAT 1954, (1898) TO END OF 1959. (12th April, 1903.)

LIABILITIES.

		In Rupees.			In English money.		
		Rps.	A.	P.	£	s.	d.
A.	Capital	5,16,320	9	1	34,421	6	8
B.	Working expenses	17,07,895	8	9	113,859	14	8
C.	Profit	6,03,308	10	3	40,240	8	0
		Rps. 28,27,524	12	1	£188,521	9	4

Assets.

D.	Block	4,10,935 12 10	27,395 13 4
E.	Cash receipts ...	17,26,261 15 3	115,084 1 4
F.	Stock in hand ...	6,90,327 0 0	46,021 16 0
		*Rps. 28,27,524 12 1	£188,501 10 8

This provisional profit of £40,240 8s. od. shows the wonderful return after six years of work of 117 per cent. on the capital expenditure of £34,421 6s. 8d., a percentage which will be materially increased when the value of the raw-silk reeled from the stock of last year's unreeled cocoons is realised.

* See also Chapter 12.

CHAPTER VI.

SILK WEAVING. CHINESE AND JAPANESE EXPORTS OF WOVEN SILK. KASHMIR TEXTILE POSSIBILITIES. PURCHASE OF LOOMS AND ENGAGEMENT OF A SILK WEAVER.

I WISH now to report preliminarily upon a suggestion I made to His Highness the Maharajah of Jammu and Kashmir in Durbar at Jammu on my arrival there.

I informed His Highness and the Durbar that I had thought for a considerable time that the cocoon-reeling results had been so satisfactory in producing raw-silk of great fineness and evenness of thread that I did not see why the natives might not now be taught how to weave such fine silks as are so largely imported into Europe and America from China and Japan, and in addition to this there is also the great field of silk-wearing India, an extensive and important market close at hand.

I stated that this idea was one of the most important that had decided me to come out to Kashmir.

I would like to say here, that for good or bad, the idea is entirely my own, and has not been suggested or mentioned to me by anyone. Being therefore solely responsible for it I have given it all the consideration in my power before I ventured to mention it either to the Government of India or to the State authorities in Kashmir.

I feel sure that I have evolved the idea of a procedure which, if thoroughly carried out, will result in a further development of the Kashmir Silk Industry almost as important as that of Sericiculture, more especially as to an excellent employment and vocation for both men, women and elder children.

I approached the Board of Trade, through my friend Sir A. E. Bateman, for information concerning the exports of woven silks from China and Japan. He kindly gave me the following statistics which show in round numbers that China exports one and a half millions sterling worth of these silks, and that Japan exports no less a sum in value than three millions sterling worth.

BOARD OF TRADE STATEMENT SHOWING THE VALUE OF SILK GOODS (DOMESTIC MANUFACTURE) EXPORTED FROM CHINA TO FOREIGN COUNTRIES IN EACH OF THE YEARS 1900, 1901, 1902 :—

Total English Values		
1900.	1901.	1902.
£1,507,358 3 9	£1,613,402 3 4	£1,333,548 14 2

STATEMENT SHOWING VALUE OF SILK GOODS (DOMESTIC MANUFACTURE) EXPORTED FROM JAPAN (EXCLUSIVE OF FORMOSA) TO FOREIGN COUNTRIES IN EACH OF THE YEARS 1900, 1901, 1902 :—

Total English Values		
1900.	1901.	1902.
£2,383,872, 8 4	£3,041,772 4 3	£3,181,667 11 10

Now these silks are imported into Europe and America because neither European nor American weavers can weave them as cheaply as the cottage-loom weavers in China and Japan. In Japan the labour was a short time ago only 3d. per day, but wages have risen there, and at the present time 6d. per day is paid for silk weaving. The maximum wages of the Kashmiris being only 4d. a day gives them an excellent chance of competing.

On my mentioning this in the Srinagar Durbar, the Maharajah rose and said he thought it was an excellent suggestion, and he should be very glad to see weaving introduced. I may mention that no silk weaving yet exists in Kashmir or Jammu. The only weaving I saw was puttoc weaving, a kind of rough woollen tweed, of which the cloaks the natives wear are made; also tweeds for tailoring purposes resembling Scotch tweeds.

I had explained that the reeling of the cocoons had been brought about, under Mr. Walton's able supervision, to such perfection, that I felt sure there was now a possibility of weaving goods quite equal in texture and fineness to those so largely exported from Japan and China, to Europe.

At the conclusion of the Durbar which lasted nearly two hours in discussing various points of the silk industry, the Maharajah again cordially shook hands with me, and thanked me in a very touching manner, and General Rajah Sir Amar Singh asked me to accompany him to his house which

PLATE 6.

Mr. Colvin the Resident, General Raja Sir Amar Singh, and the Author.

PLATE 7.

Mrs. Colvin and the Author at the Residency, Jammu, returned from an Elephant Ride to the Fort.

adjoins the Palace. He discussed the weaving idea with me, and said he felt very much interested in it, and that if I could succeed in introducing it in Kashmir, it would be nothing short of a real blessing to the people of Srinagar, many of whom were at that time without employment, and in consequence of a material advance in the price of rice, were unable to buy it.

I promised to devote my most careful attention to the subject, and I have, up to the present time, spared no pains in preparing to instal this new industry into Kashmir.

I mentioned the idea to Mr. Colvin when I met him in Srinagar a few days afterwards; he greatly approved of it, and told me to proceed on the lines suggested by Sir Amar Singh, namely, to purchase a couple of looms and engage a competent weaver to come out as soon as proper arrangements in England could be made.

I had the privilege of meeting the Church of England Missionary and Zenana Staff at Srinagar, and had much conversation with them about silk-weaving; it met with their approval. The Rev. C. E. Tyndale-Biscoe and Miss Hull told me that if a cottage-weaving industry could be established, and women could be taught to weave also, a great improvement in their material and moral condition would be effected, especially among the Hindoo widows, many of whom were living in a state of pitiable degradation.

I am not sure whether a system of cottage-weaving can be successfully instituted, owing to the great difficulties attending European supervision, the economical distribution of warp and weft, and the proper collection of the woven result; at least not in the present state of native society there. The temptation to dishonesty would be great, and the difficulties of securing an all round straightforward installation would cause much trouble, creating many drawbacks. This is Mr. Colvin's opinion, as well as Mr. Walton's.

Mr. Colvin thinks it would be much preferable to build a factory and establish, as in England, a factory system of weaving, at any rate as a first effort; and he instructed me to order twelve looms, which are now made, and sent out with all the necessary equipment for working them, namely, winding, beaming, and warping machinery.

I have done this, but in some degree a little regretfully, as I hoped to see the excellent ideas and plans of Miss Hull and the rest of the Missionary Staff have the first place, but upon further consideration I am convinced that Mr. Colvin is right, and I have been for some months engaged for the adventure, at his request and that of Sir Amar Singh.

With the help of my friend, Mr. Kershaw, of Macclesfield, who is acknowledged to be one of the most experienced manufacturers of woven silk in England, and himself one of the largest silk manufacturers, I have got out plans of an initial factory both for one and two hundred looms, capable of future extension, and showing the most economical distribution of all the necessary machinery for a hand-loom factory, and the subsidiary space for winding, warping, etc.

The consideration of power-looms must be left to the future; I think it may come in time, if found necessary,—if Chinese and Japanese competition should make it necessary.

This new Industry, if successfully installed, will find employment for great numbers of people, and it is in this way chiefly that it will be beneficial to the State. It cannot be expected to bring in the rapid and large profits of sericiculture, because of the cheap products of Japanese and Chinese hand-looms. So cheap are they, that, as I have just stated, neither hand nor power-loom work in Europe or America can yet compete with them; but I am hoping before long to evolve a power-loom which will be capable of successfully competing with their hand-loom production.

With regard to the factory versus cottage weaving, I am now all the more disposed to advocate the former, on learning from Mr. Colvin that two out of four of the new filatures for cocoon-reeling, which were almost completed when I left Srinagar in May, 1903, are now filled by women cocoon-reelers, they having been induced by pressure of hard times to leave their purdah and, perhaps for the first time in the history of industry in India, certainly in Kashmir, to come outside their homes and learn how to reel cocoons in the factory. It points to a gradual amelioration in the hard lot of the Kashmir women, who, from what I heard and saw of them, seem to bear the burden of most of the lowest and most laborious of the daily work, whilst the men take only the easiest. In this view I welcome the decision arrived at by Mr. Colvin, who recently informed me that there are now 1350 women employed.

I sent out, in March, a highly skilled weaver with the looms and the necessary apparatus; and up to that time for some months I kept him employed in using and perfecting the looms, and adapting every necessary detail, so that when he arrived in Srinagar there would be no experimental work necessary. I have already sent samples of his very successful weaving of Kashmir Silk to the Maharajah, Sir Amar Singh, and Mr. Colvin, which leave nothing to be desired as to quality. The chief point now to be ascertained is, how cheaply the Kashmiri, when his or her clever fingers be well loom trained, can compete with the Chinese and Japanese. I believe and think of it in a very hopeful way.

The woven stuffs of Japan and China are made with unthrown raw-silk from white cocoons, whilst the cocoons of Kashmir Silk are yellow and present great difficulties in overcoming the yellowness which prevents good white and pale colours from being obtained upon it.

After prolonged research and experimental laboratory work, and with the help of one of my sons,

PLATE 8.

Palace of H.H. the Maharaja of Jammu and Kashmir on the Jhelum at Srinagar.

who is an expert piece-silk dyer and finisher, we have been able so to whiten Kashmir woven Silk that it can scarcely be distinguished from the white Silk of Japan.

This is a discovery of the highest importance, and will make the progress of a weaving industry in Kashmir much less difficult of success and of sale.

CHAPTER VII.

COPIES OF PREVIOUS CORRESPONDENCE WITH GOVERNMENT, SEPTEMBER, 1889.

No. 91, Dated India Office, 26th Sept., 1889.

To His Excellency The Most Honourable the Governor-General of India in Council.

My Lord Marquis,

In continuation of previous correspondence, I forward herewith, for your Excellency's information and for such action as your Government may deem desirable, a copy of a question asked in the House of Commons on the 26th August last, by Sir Roper Lethbridge, on the subject of steps being taken in India for the improvement of Indian Sericicultue.

*2. I also forward a copy of further correspondence, as noted below, on the same subject, with Mr. T. Wardle.

3. I take this opportunity of drawing your attention to the fact that the information concerning the establishment of a Silk Committee in Calcutta,

* From Mr. Wardle, dated 12th August, 1889. From Mr. Wardle, dated 24th August, 1889. To Mr. Wardle, dated 24th September, 1889.

which appears to have reached Mr. Wardle, has not been communicated to this Office, my despatches of the 28th March, No. 39, and 4th July, No. 63, remaining unanswered.

4. I shall be glad to hear at an early date what steps have been taken, or are in contemplation, with a view to making use of the information acquired by Mr. N. G. Mukerji, in Europe, and for the establishment of the experimental Silk Laboratory for which the apparatus was sent out in December last.

I have the honour to be,

My Lord Marquis,

Your Lordship's most obedient humble Servant,

(Signed), CROSS.

Letter from Mr. Wardle to the Right Honourable the Under Secretary of State for India, Sir John Gorst.

Leek, 12th August, 1889.

Dear Sir,

When I had the pleasure of seeing you with Mr. Davenport, M.P., some time ago, you kindly suggested that I should write to you asking what had been done with regard to my suggestions to the Secretary of the Revenue and Agricultural Department, Calcutta.

In reference to the rumour of the Imperial Government having handed over the Sericicultural Laboratory and staff to the Bengal Government, who had handed it over to a mixed Committee, I thought, before troubling you, I would wait a little to see whether I should receive any reply from Sir E. C. Buck, the Secretary.

His reply has arrived, and Mr. Davenport will be good enough to hand it to you. It is just as I feared.

I am informed that the whole of this important experiment has already been handed over to three European firms, and it is not in the least likely that there will be any but an abortive result.

I deeply regret that the Government of India should not have seen the immense importance of Sericiculture in India, where the Silk Industry would have been consolidated and developed by persistent and long-continued Imperial Government direction, and an "educational" effort to improve the native rearing of silkworms.

If the Governments of France and Italy have for so many years seen the necessity of preserving their respective silk industries by State watchfulness and nurture, I feel certain that they are still more required in India.

The disease which twenty-five years ago devastated the Italian and French silkworm crops, and which has decimated, if not destroyed, the output of Cyprus, as will be seen by an admirable Report from the High Commissioner just issued to Parliament, I repeat that this disease, which twenty-five years ago did so much damage, has been effectually stamped out by the influence and work of the sericicultural laboratories, where Monsieur Pasteur's methods and discoveries are regularly in active operation.

Should these laboratories and precautions cease, pebrine and other diseases would most surely be as rife in a few years in Europe as they are at the present moment in Bengal.

Never at any period in the history of Bengal Silk was it so much in request as now. In Leek I am asked for it every day, and begged to use my best endeavours with the Government of India to do what lies in their power to enable the trade to procure this, and, if there should be the slightest doubt on this point, I would be glad to take any person you would depute to the Manufacturers of this and other English towns to see for himself that the demand for Bengal Silk is as I have represented.

The expense in India of conducting this "education" would be small compared with the annual expense in Italy and France.

The demand for Bengal Silk this year has so far exceeded the supply that Manufacturers will have to wait for the crop of the November "bund" which does not arrive in England until the early part of next year.

I cannot understand why the apparatus for the Laboratory is not required, and the question what is to be done with it completely astonishes me. If it is not the right kind of apparatus, it ought to be replaced by what is the right kind. There is no difficulty: I could easily select all proper instruments and appliances from either Lyons or by advice of Dr. Quajat at the Government Station at Padua.

I venture to think this is an Industry which the Imperial Government ought to encourage, both with regard to its value to England, and for the employment of the poor people of the silk districts in Bengal, whose loss has been and is enormous through the decay of this Industry.

Mr. Finucane's idea that the Silk experiments should be carried out by the Silk Merchants is totally in opposition to the wiser decision of the French and Italian Governments, who know too well the value and importance of the Silk Industry to their peoples.

Were I a Director of Agriculture in Bengal, I cannot help saying it seems to me, that this is one of the objects which would strongly appeal to me for Government succour and direction.

I am perfectly satisfied that the experiments will not have a lasting benefit in India unless they are placed on a permanent and scientific footing.

I take it after much concern and study of the question, that the poor peasants, who cannot help themselves nor sustain for themselves any organised system, and are mainly interested in these experiments, ought to have the direct help which I so earnestly seek for them, and I have frequently repeated that the question of producing Bengal Silk for native uses is a more important matter than that of its production for export.

Were I in Parliament I would move for a Commission on this subject. I am quite certain it is of sufficient importance for one.

I hope you will consider the deep interest I feel will be my excuse for troubling you at such length.

Yours truly,

(Signed), THOMAS WARDLE.

P.S.—The India Office note on the paragraph about Laboratory apparatus is:—"The apparatus ordered and sent out was what M. Rondot recommended, and is believed to be what (or nearly all that) Mr. Mukerji desired.

No. 3.

Leek, Staffordshire,

24th August, 1889.

Dear Sir John Gorst,

I have your letter of the 31st July before me, and cannot help expressing my deep regret and alarm at the inertness of the Government of India in the matter of the revival of Bengal sericiculture, with its possibility of great extension in the Madras Presidency, in Kashmir, and probably other parts of India, and its practical abandonment of the promising and initial steps to carry out a reformation of this Industry.

I now venture to trouble you at a little more length than before; at the same time fearing I am trespassing too much upon your valuable time.

I am quite convinced that this matter is one of such importance as to deserve a Parliamentary Committee and the evidence of experts.

I think I ought to put you in possession of some ideas, which to my mind are overwhelmingly conclusive of the absolute necessity of retaining this sericicultural teaching initiative and "education" (to use a French term) under Imperial direction.

It makes a great difference to the character of the experiments whether they are carried on by a Provincial or Imperial Institution, and especially whether they are carried on permanently or temporarily.

A permanent and not a temporary sericicultural appointment ought to be insisted upon.

It is very clear to me that the Bengal Government has no intention of carrying on sericicultural training permanently.

I may mention, in passing, that one of the most important things to be done is the translation into the Indian vernacular of the important French and Italian treatises on sericiculture.

This could only be done by a permanent and Native Official, and it will never be done under the direction of business men, especially by those who are interested in the least important side of the question, namely, in the production of raw-silk for export, neglecting entirely the silk for native uses, of which large quantities are brought from other Countries into India.

It is absolutely essential that continuous experiments be made at the expense of a special fund provided for that purpose.

Besides the benefit to India, the experience gained by such an Institution would greatly benefit our English colonies, where sericiculture is commencing.

The Silk Industry of Kashmir has up to the present time entirely collapsed from want of sericicultural knowledge, as that of Bengal is now collapsing.

Nor is it only a specific knowledge of how to grapple with the silkworm diseases which is necessary, but a special knowledge regarding the management throughout that has revived the silk Industries of France and Italy so completely.

Before the silkworm diseases appeared in France and Italy the average quantity of cocoons obtained from 1 oz. of eggs was 17 to 20 kilogrammes, but now by the sustained Government action it is over 40 kilogrammes per oz., and increases every year.

The average of 24 sericicultural departments in 1887 shows 25·99 kilogrammes from 1 oz. of eggs, and in 1888, 40·99.

I am quoting from the statistics of the "Production de la Soie en France, et à l'Etranger, of the Syndicate de l'Union des Marchands de Soie de Lyon."

I am translating this work in order to show the Government the immense importance the French attach to their silkworm breeding.

The methods that combine to produce this result in Europe at present ought to be applied wherever sericiculture exists or may be introduced, because there is little doubt that the rearing of silkworms may be gradually abandoned in Europe; the reason will be the same as the reluctance on the part of English farmers to grow wheat, and it will go back to tropical countries where human wants are few, and where labour must always remain cheap.

From the same source I quote the following remarks on sericiculture in Spain, where, from disease and other causes, this Industry has become considerably decreased :—

"M. Le Ministre de l'Agriculture, vivement préoccupé de la décadence d'une industrie, autrefois source principale de richesse pour un grand nombre de provinces, a proposé à la Reine Régente la sanction d'un décret quie institue dans quelques provinces d'Espagne une ou plusieurs stations bacologiques, qui auront pour mission de développer la culture due mûrier, produire des semences saines, et d'enseigner aux éducateurs les méthodes les plus rationelles d'élvage."

If the Minister of Agriculture of Spain considers this Industry of so much importance as to propose to the Queen Regent such an action as I am so strenuously advocating in India, I think such an argument might well have great weight with the Government of India.

The British Empire is more interested in the

prospects of tropical countries than any others, and of the making or marring of the silk trade of the Empire; the standing or falling in the world's competition will depend chiefly on the boldness with which scientific knowledge is taken advantage of, and the sooner it is done in India the better.

Science has done more to improve sericiculture than probably any other agricultural industry.

Japan has successfully taken this matter in hand, and is well ahead of us, and there is a sericicultural organization in China extremely thorough and increasing.

There ought, I think, to be a school of sericicultural training for India and for the colonies; a bureau of sericiculture ought to be established for the Empire, and it would be much better established in India.

It requires no costly establishment nor expensive laboratory instruments to start such an institution. In fact the "material" already purchased by the Government of India and sent out to Calcutta is, I believe, sufficient, or almost sufficient, for the purpose, and it appears to be not yet even unpacked. This surely is a cruel waste of opportunity and of money.

Reverting to the Government Report, I see it is arranged that the sericicultural experiments shall be managed by the Committee of Silk Merchants up to the 28th of February next, the Government paying Rs. 7,500, the remainder of the expenses being defrayed by the Committee. What is to be done afterwards?

Probably it will be said that the experiments have proved successful, and that the natives ought to be able to carry on sericiculture without Government help.

Those who know anything about the Natives of India will, I think, not expect much persistence or perseverance from them. They will be sure to give up such experimental work.

Silkworm rearing, from its very nature, must always remain in the hands of the native peasantry as a domestic industry, and differs from the industries of indigo and tea, which are carried on by the European capital, enterprise, and knowledge. Is not this a serious argument for Governmental guidance?

I am certain that, without extraneous help, and State instruction, the production of raw-silk in India cannot be improved.

In addition to this there is the question of the supply of mulberry leaf for the experimental work. Has the Committee any funds for this purpose, or are they likely to lay out a plantation if the experiments are not likely to be carried on permanently.

One of my reasons for writing to you now is that I see the Indian Budget is announced for Tuesday night, and whether it is or is not a subject for mention in the House you will be the best judge; but I am sending copies of this letter to Mr. Harry Davenport, our Leek member, who, I am sorry informs me will not be able to be present on the night of the Budget, so I also sent them to three other friends in the House, Mr. J. A. Bright, Mr. Woodall, and Mr. Kendrick, who I know wish this industry well, and would not mention it in any factious way. It seems to me to be desirable in the present stage that a question should be asked about it. It is possible that should you not care to mention it yourself you might not object to one or other of these gentlemen mentioning it. If you would prefer that it should be passed over I am sure that if you would kindly say so to them they would yield to your ruling.

Yours very truly,

(Signed), THOMAS WARDLE.

Sir John Gorst.

No. 4.

R. S., & C.,

Dated India Office,
24th Sept., 1889.

Sir,

With reference to your letters of the 12th, 24th, and 30th ultimo., to Sir John Gorst, I am directed by the Secretary of State for India in Council to inform you that no reply has, as yet, been received to the communications which, as you were advised by the letter from this Office dated 24th June last, were forwarded to the Government of India on the subject of the establishment of an experimental silk laboratory in India.

A further despatch has now been addressed to the Viceroy with a request that Viscount Cross may be informed what steps have been taken, or are in contemplation, with a view to the improvement of the Indian silk industry, and on receipt of a reply you will be again communicated with.

I am, etc.,

(Signed), Horace Walpole.

T. Wardle, Esq.

CHAPTER VIII.

No. 1.

MEMORANDUM ON THE ENGAGEMENT OF A SERICI-CULTURIST, AND THE PURCHASE OF SILK-REELING MACHINERY, EGGS, ETC., FOR THE KASHMIR STATE. STATEMENT BY SIR GEORGE BIRDWOOD, INDIA OFFICE, 12TH OCTOBER, 1897.

ON the 13th January of the present year I received a letter from Lieutenant-Colonel Sir Adelbert Talbot, K.C.I.E., the Resident in Kashmir, dated the 6th December 1896, in which my assistance was asked in obtaining the services of a scientific and practical silk expert, either a Frenchman or an Italian, to superintend the Kashmir silk industry. I was at the same time informed that Captain Chenevix-Trench, the Assistant-Resident, was proceeding to England on furlough, and had been requested to consult me with regard to the purchase of improved silk machinery and cocoons required by the State Durbar.

In February Captain Chenevix-Trench called here and discussed the matter with me, and he was at once placed in private communication with Sir Thomas

Wardle of Leek. Meanwhile the Government of India had been asked by telegraph (4th February, 1897) whether the employment of a foreigner, and the purchase of silk machinery for Kashmir had been sanctioned, and a reply (7th February) received to the effect that they had no objection to Sir Adelbert Talbot's proposals.

The matter was then (17th February), writes Sir Adelbert Talbot, officially referred to Sir Thomas Wardle, who was directed to consult the Special Assistant with regard to securing a duly qualified Italian sericulturist for Kashmir. Sir Thomas, in reply (24th February) to the India Office letter, suggested that Sir A. Talbot might, before any other steps were taken, be asked by telegraph whether Mr. C. B. Walton of Mussoorie, a gentleman he considered in every way fitted for the post, would be willing to accept the appointment. Sir Thomas Wardle also proposed that Captain Chenevix-Trench and he should be authorised to at once proceed to the Continent to purchase the silkworm eggs and silk-reeling machinery required by the Kashmir Durbar. On this the following telegram was sent (24th February) to the Viceroy: "Wardle highly recommends C. B. Walton, Montrose House, Mussoorie, for sanctioned appointment in Kashmir, and suggests, with Lawrence's approval, Talbot should see Walton before engaging foreigner." On the 16th March Sir A. Talbot informed the Government of India that Mr. C. B. Walton had been selected as sericulturist for Kashmir, and this was subsequently approved by the Government of India.

Sir Thomas Wardle on the 8th April wrote stating that Mr. Walton had asked him to purchase fifty tavelettes for Kashmir, and Sir Thomas expressed the opinion that tavelettes without the complete improved silk-reeling machinery would be useless; and he urged that three, four, or six of the best silk reeling machines it was possible to obtain on the Continent should be sent out, and that he and Captain Chenevix-Trench should be deputed to France and Italy to make the necessary purchases.

Before it was possible to authorise the purchase of the machinery the sanction of the Government of India had to be received, and the following telegram was therefore sent (15th April) to them:—

"Walton at request of Talbot asks Sir Thomas Wardle to buy and despatch silk-seed and silk-machinery, total cost about £200, besides journey to Italy. Captain Trench wishes to accompany Wardle. Does Kashmir Durbar agree to this, and to bear additional cost of his deputation?"

The Government of India replied (24th April) that the Durbar agreed to the above proposals. Sir Thomas Wardle was then (27th April) informed that his proposals of the 8th April were approved, and he was authorised to at once proceed, with Captain Chenevix-Trench, to Italy and France to buy the required improved silk machinery, and to afterwards submit a report of their journey.

In June Messrs. Wardle and Trench forwarded their joint report to this Office, together with a statement of their deputation expenses. Subsequently in September Sir Thomas Wardle submitted the complete accounts for the French and Italian machinery purchased and forwarded to Kashmir.

The sums paid by this Office on account of the purchase of the machinery for the Durbar are as follows:—

	£	s.	d.	£	s.	d.
Deputation expenses:—						
Sir Thomas Wardle ...	91	15	4			
Captain Chenevix-Trench	35	3	8			
				126	19	0
Cost of Machinery:—						
Signor Battaglia	66	19	9			
M. Berthaud	34	8	11			
				101	8	8
Or a total cost of				£228	7	8

The above amount (£228 7s. 8d.) will have to adjusted between the Government of India and the Kashmir Durbar.

A copy of the official correspondence is appended.

GEORGE BIRDWOOD.

List of Correspondence.

Sir Adelbert Talbot to Sir George Birdwood, 6th December, 1896.

India Office to Sir Thomas Wardle, 17th February, 1897, with enclosures.

Sir Thomas Wardle to India Office, 24th February, 1897.

Captain Chenevix-Trench to India Office, 24th February, 1897.

Sir Thomas Wardle to India Office, 8th April, 1897, with enclosure.

India Office to Sir Thomas Wardle, 27th April, 1897.

Sir Thomas Wardle to India Office, 3rd May, 1897.

India Office to Sir Thomas Wardle, 7th May, 1897.

Sir Thomas Wardle to India Office, 11th June, 1897, with report, &c.

India Office to Sir T. Wardle, 12th August, 1897.

India Office to Captain Chenevix-Trench, 12th 1897.

Enclosure 1 in No. 1.

Sialkot, Punjab,
6th December, 1896.

Dear Sir George Birdwood,

I had the pleasure of being introduced to you some years ago, and, although I cannot expect you to remember that, it induces me to write to you on the subject of the silk industry in Kashmir, where I am now Resident, in which I see from the correspondence forwarded with the Secretary of State's Despatch No. 51, Revenue, dated 7th May, 1896, you have taken a great interest.

The Durbar have considered proposals by Mr. Mitchell to give him and his friends a monopoly of silk-growing in Kashmir for a term of years, but prefer to keep the industry in their own hands. To encourage a revival of it as much as possible they have decided to procure the services of a scientific man, either Frenchman or Italian, on a liberal salary, with share in the profits after the sales have reached £5,000, for a term of five years. My Assistant for Leh, Captain G. Chenevix-Trench, is going to England immediately on furlough, and will consult you as to the best means of giving effect to the wishes of the Kashmir Durbar, and I should be much obliged if you would help him in endeavouring to obtain the services of a scientific but practical man, capable of directing silk operations, and especially of the microscopic observations necessary to ensure healthy seed only being used. Captain Trench knows Kashmir well, and would be able to personally explain matters to the expert, and give any information he might require. He has also been requested by the Durbar to buy such machinery of a modern type as may be necessary to secure the best results. Perhaps you would be kind enough to place him in communication with Mr. Wardle on this point. Captain Trench would telegraph to me the probable cost, and funds would be at once remitted by the Durbar through me.

It is important that the expert should reach Kashmir as soon as possible, say by the 15th March, or at any rate by the end of that month, so that he may supervise next year's operations. The supply of examined seed is unfortunately very limited, only eleven seers or so, and I am informed that it is now too late to get out more; otherwise it would have been well for the expert to buy and bring out some himself to supplement the small supply available for next year.

I am very anxious that the efforts of the Durbar to revive this industry should meet with support, and this must be my apology for troubling you.

Yours, &c.,

A. C. TALBOT.

Sir George Birdwood, K.C.I.E., C.S.I.

Enclosure 2 in No. 1.

R. & S. 99/97.

India Office,

17th February, 1897.

Sir,

I am directed by the Secretary of State for India in Council to forward for your information and guidance the enclosed copies of letters received from Sir Adelbert Talbot, Captain Chenevix-Trench, and Mr. Walter Lawrence, regarding the means proposed to be taken by the State Durbar for the development of the silk industry of Kashmir.

Lord George Hamilton will be greatly obliged by your taking measures, in consultation with Sir George Birdwood, to secure the services of a scientific Italian sericiculturist for employment under the State Durbar in the work detailed by Sir Adelbert Talbot. It is advisable that the salary of the appointment should be progressive, beginning at Rs. 500 or 600 a month, and rising by yearly increments of Rs. 50, 75 or 100, a month to a maximum of Rs. 1,000 a month. The exact amount can of course only be fixed after you have made inquiry as to the terms duly qualified experts may be willing to accept.

You will see that the person appointed to the proposed office is to have an interest in the sales of silk after they have reached £5,000 a year.

I have, &c.,

A. GODLEY.

Sir Thomas Wardle, Knt., F.C.S., &c.

Enclosure 1 in above.

Culver, Exeter,

11th January, 1896.

Dear Sir,

I am sending you a letter which Sir Adelbert Talbot, Resident in Kashmir, gave me to give to you, on the subject of sericiculture in Kashmir. I had hoped that at the same time I sent the letter I should have been able to have called on you, and obtained your valuable assistance in securing the expert which we want for the undertaking.

I am sorry to say I am quite unable to come to London for a short time, but I am asking Mr. Lawrence to kindly interest himself in the matter for me, and so make a start, for time is rather short.

Mr. Lawrence will I hope write to you, and when I am able to come to London I hope you will be able to give me an opportunity of seeing you personally.

Believe, &c.,

G. CHENEVIX-TRENCH.

Enclosure 2 in above.

22, Sloane Gardens, S.W.,

12th January, 1897.

Dear Sir George Birdwood,

My friend Captain Chenevix-Trench has asked me to forward you the enclosed. In his letter to me he says that the Kashmir State would be willing to pay Rs. 1,000 per mensem, and an interest in sales after they have reached £5,000.

Might I venture to suggest that the pay should be progressive, say starting at Rs. 500 or Rs. 600.

Yours, &c.,

WALTER LAWRENCE.

Enclosure 3 in No. 1.

Leek, Staffordshire,

24th February, 1897.

Sir,

I have the honour to reply to your letter No. 99 (R. and S.), dated 17th February, 1897, requesting me to consult Sir George Birdwood regarding the engagement of a scientific Italian sericiculturist for employment under the Kashmir State.

In accordance with these instructions, I have come up to London and have seen Sir G. Birdwood, Mr. Lawrence, and Captain Chenevix-Trench, having read the requirements stated in the letters received from Sir Adelbert Talbot, Captain Chenevix-Trench, and Mr. Walter Lawrence which you enclosed.

After discussing the matters with these gentlemen, I have come to the conclusion that the best course for the Kashmir Durbar to pursue would be for them to try to obtain the services of Mr. C. B. Walton, a few years back Superintendent of the important Silk Filature at Surdah in Bengal.

I personally know Mr. Walton, and fully believe him to be thoroughly qualified to undertake the required duties. He is well acquainted with all the varied details of Indian sericiculture. He is now residing in India, and his address is Montrose House, Mussoorie.

With regard to the second and very important part of the Resident's letter, the purchase of seed, and of modern special silk machinery for cocoon reeling, it will be necessary that I should visit Italy, particularly the Italian Governmental Sericicultural Station at Padua, and also the silk district of Milan.

I would propose that Captain Chenevix-Trench, Assistant Resident in Kashmir, now in London, should be asked to accompany me to Italy.

I understand from the Resident of Kashmir's

letter that Captain Chenevix-Trench has authority from the Durbar to purchase a limited number of cocoon-reeling machines, and 29 seers of silkworm eggs. I cannot state what the cost of the machinės would be, but I believe they are under £15 each.

Captain Chenevix-Trench informs me that he considers the Durbar would not require more than from five to ten machines at first. I would propose therefore making an arrangement for the purchase of the number required of the most modern and approved construction, in conjunction with Mr. Chenevix-Trench, and informing you by wire of the probable cost, for communication, if necessary, to the Resident of Kashmir. Similar arrangements can be made as regards the purchase of the 29 seers of eggs required by the Durbar.

I estimate the probable cost of my expenses to be as under:—

Journey to Milan and Padua £10.

Return Journey £10.

Hotel expenses at 31/6 per diem from date of departure to date of return, most probably about three weeks.

The above expenses are approximate and could be adjusted on return.

As it is most important that the very best cocoon-reeling appliances be purchased, I think it desirable to obtain the opinions of experts both in France and Italy, and as I am acquainted with several filateurs in France, I think it would be highly desirable that on our way to Italy we should visit Alais and Montpellier and see the most recent improvements in reeling in operation, for it is on this that success in Kashmir will mainly depend.

I think it would be advisable to purchase five of the best French reeling machines and five Italian ones.

In this case, the cost of travelling, say from Alais to Milan, would be increased by about £5, and the time lengthened by a few days.

Mr. Lawrence strongly advises that an additional microscope be purchased and sent out to Kashmir, as that now used is not very satisfactoıy in the detection of pebrinised moths and seed. I propose that I select a suitable instrument at Messrs. Beck's, the eminent microscope makers in Cornhill, London. I would select suitable objectives for the work they would have to do. The cost of the complete instrument would be from £10 to £12.

I have, &c.,

THOMAS WARDLE.

The Under Secretary of State
for India.

Enclosure 4 in No. 1.

1, St. James's Street,

24th February, 1897.

Sir,

Before leaving Kashmir on four-and-a-half months furlough, I was deputed by the Kashmir State to arrange with Sir Thomas Wardle, Sir George

Birdwood, and Mr. Lawrence to secure the services of an Italian sericiculturist expert.

I have seen Sir George Birdwood and Mr. Lawrence. I understand that Sir Thomas Wardle suggests that, if possible, the appointment should be given to an Englishman in preference to a foreigner.

He suggests the name of Mr. Walton, and asks that the Resident of Kashmir may be asked by telegram to personally see this gentleman, and inform us whether he would suit.

Sir Thomas Wardle has, I think, already addressed you on this point.

As it has been proposed that I should accompany Sir Thomas Wardle to Italy, to at any rate choose the required machinery and seed, I have the honour to ask that, if this proposal is approved of, I may be put on special duty from date of departure to Italy to date of return to England, and that this period may be considered as an interruption to my furlough, which now ends on the 27th April.

I estimate that the cost of the journey would be the same as that applied for by Sir Thomas Wardle, viz. :—

Journey to Padua and Milan £14.
Return ditto to England £14.
Hotel expenses at 31s. 6d. a day, from date of departure to date of return.

I have, &c.,

G. Chenevix-Trench,
Captain.

The Under Secretary of State for India.

Enclosure 5 in No. 1.

Leek, Staffordshire,

8th April, 1897.

Sir,

I beg to enclose a letter I have received from Mr. Walton, who I am glad to find has been appointed Sericiculturist to the Kashmir State.

From this letter it appears that the Resident authorises Mr. Walton to request me to purchase 50 tavelettes.

I am distinctly of opinion that the tavelettes should not be sent without the rest of the machinery, that is, without specimens of the complete machines, and I venture to suggest that three, four, or six of the best machines, containing the tavelettes which are required, and which are absolutely necessary in order to reel silk sufficiently well for European uses, should be selected by me in Italy and in France.

Considerable economies have been made in the cost of reeling, both in France and Italy, with two results (1) a diminished cost, and (2) greater excellence of quality, so much so that these methods are now being adopted both by the Chinese and Japanese extensively, and it is necessary in the interests of Kashmir that the most perfect appliances should be bought.

I think, if my suggestion is carried out, no more machines than I have mentioned would be necessary, because in Kashmir the machines could be reproduced as well as the tavelettes.

There is urgency in this matter, because the new crop of cocoons will be ready to be reeled in May, and if the reeling is done in the same slovenly way that it was done last year, the prospects of Kashmir will be seriously thrown back, and I think it is of the utmost importance that the most perfect and economical machinery be sent out at once, I mean the 50 tavelettes named by the Resident and Mr. Walton, and also the pattern machines. I do not think one pattern of each kind would be sufficient, because it would not show the connection of the bassines one with the other.

If you will authorise me to go down to France and Italy, I will do so at once, and as Captain Chenevix-Trench is very anxious to accompany me, it would probably be a very good thing for his future in Kashmir that he should do so.

Yours, &c.,

THOMAS WARDLE.

The Under Secretary of State for India.

Enclosure in above.

15th March, 1897.

Dear Sir Thomas Wardle,

Since writing to you, the Resident has been over to see me, and to inform me that Sir George Birdwood and Captain Trench propose going to Italy to buy seed and reeling machines, if my services are engaged.

The Resident has asked me to ask you to send some tavelettes by parcel post; if you would kindly do so we shall begin reeling about middle of June. They might be sent to Bombay consigned to King, King & Co., with instructions to send them at once by passenger train to the Resident or me at Srinagar, Kashmir.

I do not know how many you could procure in England, but we should require some 50 at first at any rate, if you would send as many as you can and let know the cost, &c., I would refund it to you. This year of course we can only make experiments of the reeling with Italian machinery. I think it would be the greatest thing for the industry to get a good quantity of seed, as there is no lack of mulberry and the rearers will soon take to rearing when they find so much interest being taken. It would be my first object to enlist the sympathies of the zamindars to induce the villagers to take up rearing on a large scale, and I shall constantly visit all cocoon-rearing villages to show them how they must make their huts to suit. With kindest regards.

Yours, &c.,

C. B. WALTON.

Enclosure 6 in No. 1.

India Office,

27th April, 1897.

Sir,

I am directed by the Secretary of State for

India in Council to acknowledge the receipt of your letter of the 8th instant, enclosing a letter from Mr. C. B. Walton, who has been appointed Sericiculturist to the Kashmir State from the 1st May of this year; and advising, with reference to the requisition of the letter for 50 tavelettes to be forwarded to him by parcel post, and that Captain Chenevix-Trench, Assistant Resident in Kashmir (Leh), and now at Cannes, and you should be at once authorised to proceed to Lyons and Milan to purchase these tavelettes and also four or six complete silk-reeling machines, half of which are to be French and half Italian machines, and in reply I am to inform you that your proposals are sanctioned.

You should arrange for the immediate despatch by parcel post, of 42 tavelettes to Mr. Walton, and for the subsequent despatch of two French and, two Italian complete silk-reeling machines, with one extra tavelette for each machine, in time to reach Kashmir if possible by the 1st of June next. It will probably be advisable for you to arrange with Messrs. Henry S. King & Co., 65, Cornhill, E.C., for the safe delivery of the complete reeling machines in Kashmir; they should be sent on from Bombay by passenger train. Both Captain Chenevix-Trench and yourself will be granted first class travelling expenses and a subsistence allowance at the rate of one guinea a day for the period during which you may be engaged on this deputation, which it is understood will not exceed a fortnight.

A joint report of your proceedings, together with the account of the expenses incurred, should be furnished to this Office at your earliest convenience. A copy of this letter has been communicated to Captain Chenevix-Trench for his information and guidance.

I have, &c.,

A. GODLEY.

Sir Thomas Wardle, Knt., F.C.S., &c.

Enclosure 7 in No. 1.

India Office,

27th April, 1897.

Sir,

I am directed by the Secretary of State for India in Council to forward herewith, for your information and guidance, a copy of a letter to Sir Thomas Wardle regarding the purchase in Italy of silk-reeling machines for the Kashmir State.

I have, &c.,

A. GODLEY.

Captain Chenevix-Trench, I.S.C.

Enclosure 8 in No. 1.

St. Edward Street,

Leek, Staffordshire,

3rd May, 1897.

Sir,

I have the honour to acknowledge the receipt

of your letter of the 27th April, No. 1342/97, requesting me to proceed with Captain Chenevix-Trench to France and Italy to purchase silk-reeling machinery on behalf of the Kashmir Government.

In reply, I beg to state that I shall start on the journey on Friday or Monday next.

I am very closely engaged in my business just now, and I must endeavour to compress my journey within the shortest possible time, but you may rely on my not hurrying it over to the detriment of the duty entrusted to me.

I shall at once obtain the 50 tavelettes, but as regards the reeling machines there is a difficulty. I have had much correspondence on the subject since I last saw Sir George Birdwood at the India Office, and find that these machines are generally sold to manufacturers in batteries of six, and that to ask for two separately raises suspicions of the purchaser having some sinister intention. I desire, therefore, to be allowed discretion on this point, and if necessary to purchase a battery and a half of machines, *i.e.*, nine instead of four.

I see my subsistence allowance has been fixed at one pound per day, the same as Captain Chenevix-Trench. In doing this it must have been overlooked that I am not in the pay of the Indian Government, but engaged in an independent lucrative business of my own, which I cannot quit without incurring far greater sacrifice than even the most liberal subsistence allowance could possibly repay. Among business people such services as I am now rendering to the India Office would be paid for, beyond out of pocket expenses incurred, by an honorarium, but considering the relations on which I have always stood with the India Office, I say nothing of that, and only refer to the matter at all in the frankness engendered by my past friendly relations with the India Office, and the gratifying way in which they have recently been acknowledged. But I cannot accept a subsistence allowance on which I could not even barely subsist, and, if I am paid at all, must claim two guineas a day, or whatever may be the highest sum allowed by the India Office to specialists not in their regular employ.

My address at Lyons will be the Grand Hotel, Bellecour, and at Milan the Grand Hotel de Milan, and at Padua, care of Cavaliere E. Verson, the Director of the R. Stazione Bacologica, Padua.

I have, &c.

THOMAS WARDLE.

The Under Secretary of State
for India.

Enclosure 9 in No. 1.

India Office,
7th May, 1897.

Sir,

I am directed by the Secretary of State for India in Council to acknowledge the receipt of your letter of the 3rd instant, and in reply to inform you that your proposal to purchase nine reeling machines, if necessary, for the Kashmir State instead of four, is sanctioned, and that your subsistence allowance during the time you are making these purchases in the South of France and Italy will be, not one guinea per day, as stated in my letter of the 27th April, but two guineas per day.

The present sanction for five additional reeling machines is given subject to the condition that the total outlay, for the purchase of machinery and seed, shall not exceed £200 in all, which sum has been sanctioned by the Government of India.

I am, &c.,

A. GODLEY.

Sir Thomas Wardle, Knt., F.C.S., &c.

Enclosure 10 in No. 1.

Leek, Staffordshire,

11th June, 1897.

Sir,

Under instructions from you in your letter dated 27th April, 1897, R. and S. 1266, I have visited Italy and France with Captain Chenevix-Trench, Assistant Resident in Kashmir, and have the pleasure to send to you herewith, first, our joint Report of our journey, investigations and purchases.

Second. Collection of Italian and French cocoons described in the Report.

Third. A box containing the price-lists, diagrams, photograghs and accounts.

Fourth. A book on Sericiculture by Cavalieres Verson and Quajat.

The three last are being sent off to-day by railway, addressed to you at the India Office.

You will find full instructions as to payments of accounts.

I have drawn up the Report and have submitted the draft of it to Captain Chenevix-Trench, who has approved of it, and will sign it at the India Office.

I enclose a copy of a letter I have to-day received from Signor Giovanni Battaglia, whose promptitude and execution of the order is quite satisfactory.

Please to send a cheque to him for 340 francs as per invoice enclosed. The invoice is simply for tavelettes and not for the reeling machines complete.

I have, &c.,

THOMAS WARDLE.

The Under Secretary of State
for India.

CHAPTER IX.

VISIT TO ITALY.

Enclosure 1 in above.

Report of a journey to Italy and France on Government of India Sericicultural business, in May, 1897, by Sir Thomas Wardle, of Leek, and Captain Chenevix-Trench, Assistant Resident at Kashmir.

T. Wardle.

On arriving at Milan, I met by appointment Captain Chenevix-Trench, Assistant Resident at Kashmir, who had been staying at Cannes, and and who came on to Milan to meet me.

T. Wardle and Captain Trench.

We commenced our work in Milan by an inspection of the cocoons of the best merchants of that commodity.

Owing to the lateness of receiving a reply from Kashmir in respect of the visit to the south of Europe, we found it was quite impracticable to attempt to give orders for eggs to be of any service for this season. We decided therefore not to purchase any. We were further confirmed in our view that this was the right course from a telegram received 9th May from Mr. C. B. Walton from Srinagar, as follows:—

"At the Resident's request I am wiring Bombay to advise that no new European seed should be sent out till October."

We thought it would be much the best to defer the purchase of seed until samples of cocoons had been selected and sent out to Kashmir. Such a collection, which has been most carefully selected, accompanies my Report.

The houses we visited at Milan were the following:—

No. 1. Rinaldo et Agostino Casati, via St. Andrea 19.
No. 2. Signor Susani, via St Andrea 2.
No. 3. Guillo Cesari Albini, via Manzoni 43.
No. 4. Biffi di Filippo, via Cusani 14.
No. 5. Signor Numa Laval, via Broletto 37.
No. 6. Cav. Laugier Félix, via Durini 23.

The following is a seriatim account of our visits to these houses.

1. Rinaldo et Agostino Casati.

This is an excellent house. The prices of their seed ("graine, eggs") were from 8 lira per ounce of 30 grammes. We obtained from them nine varieties of cocoons, of which we send you samples of each as follows:—

No. 1. Varo.
No. 2. Ascoli.
No. 3. Brianza.

No. 4. Incrociata Giallo. Marca Sublime.
No. 5. Incrociata Bianco Giapp. Con. Giallo.
No. 6. Incrociata Shanghay Sfer. Con. Giallo.
No. 7. Incrociata Chinos Oro. Con Giallo.
No. 8. Chinese. Marca Oro.
No. 9. Bianco Giapponese.
No. 10. Bianco Shanghay.

From these cocoons it will not be difficult for Mr. Walton to select seed when purchasing.

We would most distinctly recommend some purchases to be made from this selection, in fact we think it would be well if the amount of seed required to be purchased for next year's crop could be divided amongst each of the dealers we are reporting upon.

In this way cropping results could be compared simultaneously and a reliable opinion arrived at in one year.

We distinctly promised Signor Casati he should have an order, and the cocoons he has given us are, in his opinion, the best for exportation to Kashmir. We send herewith his book of instructions and price list. (D, E.)

2. Signor Susani (Istituto Bacologica Susani).

This is a well known house, and the one from which Mr. Mukerji purchased eggs in 1892 at 15 francs per ounce.

He deals in two kinds of eggs; those he calls "cellulare" are those which have had all the moths microscopically examined; the eggs are guaranteed free from pebrine corpuscles, and those which he calls "industrials" are those in which the moths have not been so examined. There is a difference of about one-third in the price. We enclose a copy of his list. (F, G.)

The following is a list of his cocoons:—

No. 1. Giallo Puro.
No. 2. Giallo Puro. Var.
No. 3. Bianco Giapponese.
No. 4. Bianco Indigeno.
No. 5. Bianco Chinese.
No. 6. Giallo Puro.

He recommended for Kashmir two kinds. The Bianco-Chinese and the Giallo Puro (*see* Nos. 5 and 6 in the box), the Chinese being originally from Chinese seed but reared in Italy and the Giallo Puro being the pure Italian yellow breed. He says these two breeds crossed produce excellent silk, and he thinks would best suit Kashmir. His prices are:—

12	lira	per	ounce for	50 oz.	orders.
11	"		"	200	"
11	"		"	500	"
10	"		"	1,000	"

He thinks it would be best to send out the eggs in October or November.

He said the yield from one gramme of graine of his Giallo Puro cocoons was 2 to 2½ kilogrammes of fresh cocoons, one gramme of graine of the white kind produces 2 kilogrammes of fresh cocoons, and he considers the average weight of dry cocoons is one-third that of fresh cocoons, that is 3 kilogrammes of fresh yield 1 kilogramme of dry.

This I (Thomas Wardle) have always understood to be the more correct estimate, and I know it is usual so to consider it in France.

Twelve kilogrammes of fresh cocoons of his yellow kind produce one kilogramme of silk, and sometimes 11 kilogrammes will produce the same weight.

We send you a selection of Susani's cocoons, strongly recommending them. Also a circular of his. He said he could procure any appliances for sericiculture.

3. Guillo Cesari Albini.

This house appears to be a very good one and some seed should be ordered from it, as we made a partial promise that this would be the case. He stated that the average weight of fresh (that is, undried) cocoons, obtainable from one gramme of eggs of the race he recommended, would be from 2 to 3 kilogrammes, and he also said that one ounce of graine produces 60 to 70 kilogrammes of fresh cocoons, and that the average weight of dry cocoons is one-fourth that of fresh, that is, 4 kilogrammes of fresh cocoons are equal to one kilogramme of dry.

He stated that one gramme of graine produced half a kilogramme of dry cocoons, but this I (T. Wardle) think is rather under the mark.

He stated that 8 kilogrammes of his cocoons produce one kilogramme of grège or raw-silk, with 2 or 3 ounces of "waste," some qualities of cocoons producing as much as 8 ounces of waste.

His price was 12 lira per ounce of 30 grammes, sold in 30 gramme packets, for quantities below 50 packets, but for 50 packets the price would be 10 lira. In practice the ounce contains 30 grammes, although in Italy the ounce is from 27 to 28 grammes. The kilogramme consists of 33 and one-third grammes.

He thinks the eggs should be sent out at the beginning of November, and he could not well get them ready before.

The temperature at which eggs must be kept in transit must be 17° to 18° R. *i.e.*, 62° to 64° Fahr.

He had no samples of cocoons to furnish us with.

4. Signor Biffi di Filippo.

This was the next house we visited and we enclose his price list.

We were very favourably impressed with Mr. Biffi. The following is his list:—

No. 1. Ascoli, used for crossing with White Chinese, Ascoli, and Pyrenees give the same results.

No. 2. Result of crossing the yellow Pyrenees with White Chinese.

No. 3. The Chinese White used to cross with Yellow Pyrenees and Ascoli both giving the same results.

No. 4. Yellow Pyrenees used for crossing with White Chinese; Ascoli and Pyrenees give the same results with White Chinese.

He crosses Ascoli cocoons with White Chinese. He also crosses Yellow Pyrenees with White Chinese.

The Pyrenees and the Ascoli when crossed with the Chinese give the same result.

These he recommends for Kashmir, and thinks they will do admirably.

His prices range from 14 lira per ounce to 12 lira, as will be seen from the enclosed list, No. 5. (H, I, J.)

5. Cav. Numa Laval.

We had a good reception at Cav. Numa Laval's, he evidently does a large business.

His assortment of choice cocoons consists of the following:—

No. 1. Varo.
No. 2. Var Liban.
No. 3. Giallo Pirenei.
No. 4. Giallo Pirenei.
No. 5. French Var.
No. 6. China White.
No. 7. Produced by crossing 5 and 6.

The eggs supplied by him would be ready to be sent to India in August, and could be sent out any time after that. His prices are remarkably low. They are at the present time, for 50 ounces, 7 lire per ounce, 500 ounces to 1,000 ounces, 6 lire per ounce. Of these eggs he says about 2 per cent. of them contain pebrine corpuscles.

If the order comes later than August, the price will be one lire per ounce in excess of this.

We enclose his published list, No. 6. (K).

Ten kilogrammes of fresh cocoons will produce one kilogramme of raw-silk, with 5 per cent. of waste.

He recommended the French race, but thinks also that the hybrid race crossed with French (Var) and Chinese will give the most silk, but the Chinese are always more or less infected with pebrine, and it does not signify whether it is the masculine French, or the feminine Chinese, or the reverse. The eggs come from China on cards, and have the drawback of being infected with pebrine.

6. Cav. Félix Laugier.

This is also another good house. The eggs would be ready to be sent in August. They should not be sent in ice. The price is 10 francs per ounce of 30 grammes for orders up to 100 ounces. For orders of 100/200 ounces 9 francs per ounce, and for orders of 200/500 ounces 8 francs per ounce. (L.)

One kilogramme of his fresh cocoons produces 100 grammes of silk, one-third of dried cocoons makes the same quantity.

He says that one ounce of his eggs produced 90 kilogrammes of fresh cocoons, and 30 of dry, and the average weight of fresh cocoons from one gramme of "graine" (eggs) is 3 kilogrammes, and of dry cocoons 1 kilogramme per gramme.

He thinks that the eggs ought to be sent from Marseilles about the 15th August, to arrive in Kashmir a month afterwards. On arrival they must be carried at a suitable temperature until November from zero to 15° R., from November to March at a temperature of from zero to the natural, and from March to April from zero gradually to 7°. centigrade (48° Fahr.) until they hatch.

The surrounding temperature must never be allowed to go beyond this.

He informed us that he sends out the eggs in wooden boxes full of holes in divisions, covered with zinc. He has already sent cocoons to Kashmir. It is probable that they were sent by him two years ago.

The kind he sent are of the Italian race crossed with French, similar to those we send in box No. 7, and, in his opinion, produce the best results. He seems to know all particulars about silkworm breeding, and we have no doubt is most reliable.

We should decidedly advise that some cocoons be bought from him, because he is a Frenchman whose principal works are at Cotignac (Var),

France, and the seed would probably be sent from France. We think it would be very desirable to introduce into Kashmir the French races, which, as far as we have been able to judge, are on the whole a little superior to those of Italy.

In Italy the magnanerie, that is, the breeding of worms, is entirely now a cottage industry.

In feeding the silkworms from the early to the last stages perforated paper is used, commencing with very small perforations, and increasing at each moult, through which the silkworms creep up to the food.

The holes in the first stage are not more than the sixteenth of an inch in diameter, and in the last stage at least a quarter of an inch.

Upon this perforated paper are placed fresh mulberry leaves and the worms creep through the holes to the leaves, thus securing cleanly feeding.

Articles for sericiculture can be obtained from the following houses:—

E. Marcani & Co., Via Aparti 14, Cremona.

Ditta Onorio Gavziero, Via Santa Enfernia 16, Verona.

Cav. Pietro Motta, Campocroce di Mogliano, Veneto.

In addition to the houses we have just mentioned, we would also especially recommend some to be ordered from Monsieur Laurent de l'Arbousset,

Rue Mandajors 22, Alais (Gard), Frence. He is a great expert, a silk breeder, a large distributor of seed, and a silk reeler, and almost the best cocoons we have seen have been at his establishment. In reference thereto we only need refer to the samples which will be found in box No. 8, the description of which is as follows:—

No. 1. Race à Papillons noirs, l'ancienne race pure de Cevennes.

No. 2. Race Arbousset; race jaune croisée representant, en beau, tous les principaux cocons à croisement elévés en France.

No. 3. Race ordinaire du Var, le race pure (sans croisement).

No. 4. Race Var.

These are the principal races which best represent the majority of good cocoons produced in France.

His price of graine (eggs) is 130 francs per kilogramme or 3/- per oz.=4 francs 30 centimes per oz.

In each kilogramme there are 33 ounces of 30 and one-third grammes (French).

Monsieur l'Arbousset says cocoons for India should be sent out from Alais in August or September, and the eggs ought to be kept in a mountainous and fresh district until the following spring. It would neither be safe nor practicable to pack them in ice, as they would miss the air and perish going out. The voyage should be made before they are exposed to the frosts of Europe, and it is owing to this rule not being observed that so many have been unsuccessful.

Monsieur l'Arbousset is the editor and proprietor of the "Bulletin Sericole Français," a weekly periodical, the subscription to which is 10 francs per annum, which we would strongly recommend to be regularly sent to Kashmir, as it contains much of sericicultural and silk trade interest.

The average weight of fresh ("frais") undried cocoons obtained from one gramme of "graine" is 1 kilogramme 500 grammes, and the average weight of dry cocoons obtained from one gramme of "graine" is 500 grammes.

The average weight of fresh cocoons obtained from one ounce (of 30 grammes) of "graine" is 45 kilogrammes, and the average weight of dry cocoons obtained from one ounce of "graine" is 15 kilogrammes. It takes 3 kilogrammes of fresh cocoons to make one of dry cocoons.

The average weight of raw silk (grège) to be obtained from one kilogramme of fresh cocoons is about ·085 grammes, and of dry cocoons about 0·255 grammes.

The amount of raw silk that can be reeled per day with a battery of three reeling bassines, and with three fileuses and one batteuse, ought to be for a "titre ordinaire" (11 to 13 deniers) about 1 kilogramme 200 grammes, and with good cocoons 300 grammes each fileuse more.

In addition to the Alais (Gard) cocoons of Monsieur l'Arbousset's, I (Thomas Wardle) now wish to draw attention to those with which I have been favoured by the kindness of Monsieur F. Lambert, Director of the Government Sericicultural Station, Montpellier.

I send you ten of the principal French varieties which I have received from him, with the following full description of each:—

No. 1. Pierced yellow cocoons from the Alps.

No. 2. Yellow cocoons from the Alps.

No. 3. Yellow cocoons from the Cevennes.

No. 4. Yellow cocoons from Corsica.

No. 5. Pierced yellow cocoons from the Cevennes.

No. 6. Pierced yellow cocoons from Var.

No. 7. Pierced yellow cocoons from Roussillon (Perpignan).

No. 8. Pierced white cocoons from the Cevennes.

No. 9. Pierced yellow cocoons from Heroult (Sericicultural Station).

No. 10. Pierced white cocoons from the Alps.

We think this list certainly exhausts the best dealers in eggs. We would recommend that none be purchased except those of moths which have been microscopically examined under the system of Pasteur, that is, the "Cellulare" as distinguished from the "Industriale."

CHAPTER X.

VISIT TO A COÇOON-REELING FILATURE NEAR LUINO. PURCHASE OF COCOON-REELING MACHINERY. STEAM-POWER VERSUS "FILATEUR DOMESTIQUE" REELING. TAVELLETTES. SIGNOR A. GRANZELLA.

BY recommendation of several of the best filateurs in Italy we thought it necessary that we should visit the establishment of Signor Giovanni Battaglia of Luino, on Lago Maggiore. We spent an exhaustive day there, inspecting machinery in various stages for reeling cocoons.

The system now adopted is that of the tavelette, similar to that I (T. Wardle) took out to India in 1885, but considerably modified and simplified in construction.

Before ordering at this house, we went some distance to a very large silk-reeling factory where between 200 and 300 workpeople were employed, chiefly in reeling cocoons with machinery constructed by Battaglia.

This machinery was working with the greatest possible regularity and efficiency, and we consider the journey amply successful if no other constructor had been visited.

The girls and women were doing beautiful work, turning out raw-silk of the very best quality. The wages of this factory were 50 to 80 centimes per diem for women and girls, and for children 1 franc to 1½ francs per week, all making twelve hours a day.

There was a considerable quantity of throwing machinery at work, making organzine and tram, and also a little weaving going on of cheap cotton and silk cloths for America.

We returned to Battaglia and there completed our purchases, bringing away with us working drawings of the machinery, which we send, and duplicates of which Captain Trench possesses.

The enclosed account will explain the extent of the machinery bought. It consisted of two tables, each comprising two bassines and one batteuse, with accessories to provide for breakages, altogether amounting to 1,200 francs. (M.)

We found one fileuse attends to each bassine, each bassine having over it five tavelettes, that is, reeling five compound threads at the same time, in Italy. In France four threads only were being used.

One bassine reeling four sets of cocoons in France produces 400 grammes of raw-silk of the ordinary "titre" (11 to 13 deniers) per day, worked by steam power. Battaglia's five tavelettes yield 500 grammes per day. One batteuse prepares cocoons for two bassines.

Besides these, we ordered 30 tavelettes complete, which were to be sent off in ten days at furthest by Parcel Post to Mr. C. B. Walton. The total bill amounts to 1,690 francs, 27 centimes, payment of one-third of which was demanded on giving the order, the rest to be paid for when the machinery is ready. Captain Trench undertook to send a cheque for this one-third, and in the course of at furthest two months, an application will come to you through us, for the remaining two-thirds.

It was necessary to purchase these three machines of a construction to be heated by a wood-fire, with furnace underneath, because it was understood from Captain Trench that there was no steam power arrangement for cocoon-reeling in Kashmir. These machines will be so constructed as to be worked together by one turner, or separately. One young person can easily turn 20 of them, but in the factory we visited there were about 50 similar machines in each row, and all turned by steam power, the water of the batteuse and the bassines being also heated by steam, and all working with great regularity.

We should advocate the adoption of this steam power system in Kashmir over and above the other method, which they term "filateur domestique." The Plan 2074 shows how the bassines (in which the cocoons are at first softened and their reelable ends found by a brush) are heated by a "foyer" underneath.

The important part of this machine is the "croissure" where the threads of four cocoons are twisted round each other and made to pass over three small glass bobbins or pulleys before the compound thread goes on to the reel at "a." This pulls all the kinks out of the thread and causes it to be conveyed to the reel with the least possible "duvet." The Plan 2069 gives the side view of the machinery.

In addition to the working drawings we also enclose two photographs, one of which shows two smaller tables with a separate batteuse, and another the most approved system and such as is now used most successfully in a very large filature at Covia near Luino. We also send a third photograph showing the batteuse.

Signor Battaglia was instructed in about the following words as to the forwarding of the machinery, as advised by Messrs. Henry S. King & Co. It was to be despatched to Signor E. Dionisi, of Brindisi, and to be at that port at least three days before the ship sailed, but we instructed Battaglia previously to enter into communication with Signor Dionisi before the goods were sent there. He was also told that the goods must not be exceptionally heavy, but must be packed in several cases in order that shipment might not be refused if the steamer were late in loading

8

them, so we do not think there will be any difficulty in the matter.

From Luino we journeyed by Lugano and Como to Padua, which is near Venice, and on arriving there we were most agreeably received by Cavalieres Verson and Quajat, the Government Directors of the R. Stazione Bacologica, who spared themselves no trouble in explaining to us every possible detail of the course of instruction in sericiculture at that interesting and excellent school. We saw some students in attendance at lecture, being instructed in the use of the microscope, &c., others superintending the hatching-out and feeding of the worms in separate chambers.

In each chamber there were two tables about seven feet long, each three to four feet wide, covered with cut and sliced mulberry leaves, upon which numerous silkworms were feeding.

These tables were so constructed between two vertical supports as to be capable of being inclined towards the light or being kept flat.

Another method was for each pair of uprights to contain several flat tables up to about six feet high and about 18 inches apart. Upon these tables immense numbers of silkworms were feeding on mulberry leaves which had been cut by a machine, in which the leaves were closely packed, and cut with a knife, after the manner of cutting hay, into salad-like shreds. These shreds were placed upon the perforated card paper mentioned previously.

Another method by which the worms were fed upon uncut leaves was arranged by placing branches of the mulberry tree with the leaves attached, and placed upon sloping frames near to the ground. The worms seemed to thrive well by this method. Perforated cardboard was not used in this method as the excrementitious matter falling to the ground made it unnecessary to use a perforated cardboard.

In one of the rooms the hatching of the worms had been expedited by artificial heat, and the worms were at the end of their fourth moult, and were mounting the "bruyère," many of them having almost formed their cocoons.

I (T. Wardle) would also mention the recent work entitled "Il Filugello e l'arte Sericola by E. Verson and E. Quajat, Direttore e Vice-Direttore, Della R. Stazione Bacologica Sperimentale di Padova, 2 volumes, price 10 lira each." An exhaustive and recent treatise on Sericiculture, which I would recommend to be translated into the vernacular and also into English. I bought two copies, one of which I send, the other I would like to retain, at least for perusal. It is the most recent book on the subject. (Q.)

Nothing can exceed the perfection of the arrangements of this place, and it is not surprising that the result of such excellent instruction causes

Italy to occupy such a high position in the sericiculture of the world.

About 30 species and varieties of the mulberry were being grown in the Government Gardens, as well as the ailanthus and other silkworm feeding trees.

Captain Trench took many notes, and made some drawings of the educational machinery.

We had the pleasure of meeting there with Signor A. Granzella, a young man of about 30, who was taking lessons from these gentlemen, and who had come over from China, where he was attached to the Imperial Maritime Customs, Shanghai, under Sir Robert Hart. From Padua he was going to Como to the weaving school there.

He was exceedingly attentive to us. He speaks English very well, and, being in the Anglo-Chinese Civil Service, was very much interested in our visit. His home address is Pianello, Lake of Como, where he will be up to the end of next year, but he will stay at Padua till the end of next June. His address in China is, Imperial Maritime Customs, Shanghai. He might be of service to Kashmir before his return.

He gave me (T. Wardle) some live Chinese Tussur Cocoons, with which I hope to make some experiments by feeding the worms on the English oak. I have already recommended to the Secretary of State for India that it would be very desirable to see if this species of Tussur silk could not be cultivated on oak trees on the Terai at Kashmir, being a silk at the present time very much in demand.

We went on to Lyons, France, and visited the establishment of Messrs. J. Berthaud et Fils, the principal makers of cocoon-reeling machinery.

Here we found a good deal of machinery in construction, of the most approved French type and also of the Italian tavelette system. The difference between these systems is that the "Chambon" is without tavelettes. We found opinions varying from those in Italy. Monsieur Berthaud had the conviction that for economy and good work the French Chambon system was preferable to that of the Italian tavelette; but on our pointing out to him that he was executing a considerable order for machinery in the Italian system, and on asking him if the order was for France, and he replied "yes," we said, "How is it that any French filateurs can prefer the Italian system if the Chambon is better?" He smiled and said, "Well, there are some filateurs in France who think the Italian tavelette is better and prefer it."

On examination of his machinery of the Italian system, we found little or no practical difference between his and what we had ordered from Battaglia, and we thought it better not to purchase any there, but to content ourselves with simply ordering one table of the French system Chambon.

It seemed to us there was no good in spending more money in these small hand machines ("filature domestique") when there was a possibility of Kashmir in the near future using steam-power machines.

We therefore simply ordered two bassines, with supplements for replacing breakages, altogether amounting to 710 francs. We enclose the account. With regard to payment, Monsieur Berthaud said he would be glad to receive the money whenever it was convenient to the India Office to pay it. (N.)

He will send photographs* of the machinery as soon as they have had some taken, which we will forward. This is a first-rate house. The machinery will be manufactured with the least possible delay, and will be sent to Marseilles through Messrs. Henry S. King & Co., to be forwarded to Mr. C. B. Walton, The Residency, Srinagar, Kashmir.

Attached to the machinery of Monsieur Berthaud's is a system called "Jette-bous." To one of the bassines the jette-bous system will be attached. The plan of the machinery belonging to it is shown in the accompanying diagram No. 3. Near to the eyelet-hole through which the four threads from the four cocoons pass, there is a revolving brass spindle which catches up the ends quickly and enables an inefficient worker to reel cocoons easily.

* Since received and included.

Enclosed with these papers of Monsieur Berthaud's is one which gives an account for an experimental bassine, suitable for a State or Technical School, and also having a serimetre-dynamometre for testing the strength and elasticity of the bave of the cocoon.

This machine is necessary where students have to be taught, but we did not purchase one. We thought it might be useful to Kashmir to be in possession of their description and cost. They are similar to those supplied to the R. Stazione Bacologica, Padua.

There are also a few inexpensive machines which it is recommended might be bought from Messrs. Berthaud et Fils such as "Eprouvettes" to measure the skeins before weighing and testing them. These could be ordered direct from Kashmir when wanted. The Serimetre, Fig 4, p. 2, would be particularly useful and also the other accessories in the list.

Messrs. Berthaud et Cie consider the reeling is not as good by fire-heating as by steam, because the heat cannot be kept up with regularity, and with fire-heating there is dust, which causes the silk to be less lustrous.

For Bengal cocoons the Italian system is better than the Chambon, but Monsieur Berthaud was of opinion that for Kashmir the Chambon system is better than the Italian.

With regard to the jette-bous he said it was more convenient to use it by those who are not well acquainted with the system Chambon, and the jette-bous can be used or not; but if the workpeople are to be well trained it is better not to begin with it.

The best filateurs in France use the system Chambon because it gives the best results, the thread being better and stronger than the Italian system; but the Italian produces more. Still, opinion is divided on this point.

The Chambon makes less "duvet" than the Italian because the croissure is more energetic by being at a wider angle.

The two machines of Messrs. J. Berthaud et Cie. will be sent off on the 15th June, the tavelettes in ten days by parcel post.

I (T. Wardle) have also received from Alais particulars of a cocoon-reeling machine, called "filature Antoine," which is recommended by Monsieur Laurent de l'Arbousset. I enclose his letter and one from Léon Malméjean, the constructeur, of Alais, a mechanician employed by Monsieur Laurent de l'Arbousset. The drawing is too rudely done.

As I have not seen the machine at work I did not think it well to order one, especially as we have been most fortunate in securing machines of undoubtedly the best Italian and French systems, leaving little, if anything, to be desired as to their capabilities of working in the most modern and economical way.

This completes our Report, which we hope will be found useful to Kashmir as well as India. It contains information and references of a most thorough and useful kind, and we shall be glad to learn that it gives satisfaction to you and to the authorities in Kashmir.

We have been very anxious to complete it in the least possible space of time, and from the extent of ground travelled over, we have reason to hope that you will think that as much has been done as could possibly be in the time.

We have, &c.,

THOMAS WARDLE.

G. CHENEVIX-TRENCH, Captain.

Leek, 11th June, 1897.

Illustrations, &c., to Report by Sir Thomas Wardle and Captain Chenevix-Trench.

LIST.

1. Report by Sir Thomas Wardle and Captain Chenevix-Trench.
2. Cocoons from Casati, two boxes, Nos. 1 and 2.

Cocoons from Susani, one box, No. 3.

" " Biffi, one box, No. 4.

" " Numa Laval, two boxes, Nos. 5 and 6.

Cocoons from Laugier, one box, No. 7.
„ „ Arbousset, one box, No. 8.
„ „ Lambert, one box, No. 9.

A.B.C. Three photographs of silk reeling machinery from Battaglia.

Three diagrams of cocoon reeling machinery from Battaglia, Nos. 2069, 2074, 2134.

D.E. Instruction book and price list of Casati's.

F.G. Prospectus and price list of Susani's.

H.I.J. Prospectus, price list, and order sheet of Biffi di Filippo.

K. Prospectuses from Numa Laval.

L. Instructions and price list of Felix Laugier.

M. Account of Battaglia's amounting to 1690 francs 20 c.

N. Berthaud's accounts amounting to 710 francs.

O. Price list of J. Berthaud et Fils.

P. Berthaud's estimate amounting to 510 francs.

Q. Diagram of Jette-bous from Berthaud.

Book, one volume, "Il Filugello e l'arte Sericola," by E. Verson and E. Quajat.

R. Letter from Monsieur Laurent de l'Arbousset.

S. Estimate from Léon Malméjean.

Enclosure 2 in above.

Luino, 8th June 1897.

I received your letter of 28th May. We are hurrying on with the construction of the apparatus for "croissure" (tavelettes), but some time has been lost on account of the postal arrangements. This is what has happened: The Luino post office, on being asked if there was any parcel post connection with Kashmir, answered in the affirmative, and when the boxes were sent there, they would not take them in, saying that they had made a mistake and that there was no parcel post arrangement with Kashmir. I then applied to the Swiss post office and have succeeded in sending off the six boxes by Monteggio (Canton Tessin). They were sent on the 1st instant, and I am now enclosing the bill for 340 francs.

I am sending one copy to Captain G. J. Chenevix-Trench and another to Mr. H. Walton at Srinagar.

With regard to the first parcel, I must tell you that in the demand 2146 of 13th May, we made a mistake in your favour charging too much for the tavelettes. We invoiced them at 330 francs, whilst with carriage, packing boxes, and exchanges, this sum is only exceeded by 10 francs.

Notice that 76 francs are for carriage and 12 francs for packing cases.

I am sending you duplicates of the designs 2069, 2074, and 2134, and six photographs.

I remain, &c.,

GIOVANNI BATTAGLIA.

Sir Thomas Wardle.

One bill, six photographs, six designs, together.

Enclosure 11 in No. 1.

R. & S. 1708/97. India Office,

Sir, 12th August, 1897.

I am directed by the Secretary of State for India to acknowledge the receipt of your letters of the 11th and 18th June last respectively, submitting the report prepared by Captain Chenevix-Trench and you on your recent mission to the Continent, for the purchase of silk machinery for the Kashmir Durbar, and a statement of the travelling expenses incurred by you on the occasion. In reply I am to inform you that the Accountant-General has been instructed to forward you a draft for £91 15s. 4d. on the latter account, and to pay any duly certified bills up to a total of £200 on the former account. The account for machinery should be countersigned by Captain Chenevix-Trench and yourself.

On hearing from you that the whole of the machinery, &c., has be supplied, a copy of your report, with accompanying samples, will be forwarded to the Kashmir Durbar.

I have, &c.,

A. N. Wollaston.

Sir Thomas Wardle, K.B., F.C.S., &c.

Enclosure 12 in No. 1.

India Office,

Sir, 12th August 1897.

I am directed by the Secretary of State for

India to forward you the enclosed copy of a letter addressed to Sir T. Wardle regarding your joint report on your recent deputation to the Continent to purchase silk machinery on behalf of the Kashmir Durbar, and with reference to your claim for £51 19s. 8d. for travelling allowance, &c., to inform you that the subsistence allowance admissible to you by rule is at the rate of one guinea a day. The Accountant-General has therefore been instructed to forward to you a draft for £35 3s. 8d.

I have, &c.,

A. N. Wollaston.

Captain Chenevix-Trench, I.S.C.

CHAPTER XI.

REPORT OF AN INFORMAL MEETING HELD AT THE INDIA OFFICE ON THURSDAY, THE 26TH SEPTEMBER, 1901, TO DISCUSS THE PRESENT POSITION AND FUTURE PROSPERITY OF THE NEW SILK INDUSTRY OF KASHMIR.

PRESENT: Sir Thomas Wardle, in the Chair; Sir George Birdwood, K.C.I.E., M.D., LL.D., C.S.I.; T. W. Holderness, C.S.I., I.C.S.; Sir Adelbert Talbot, K.C.I.E.; L. W. Dane, C.I.E.; and two other gentlemen connected with silk and sericiculture in London and Calcutta.*

Sir George Birdwood said he had arranged this informal Meeting in order to give Sir Thomas Wardle and the other gentlemen present, all of whom were interested in the revival of the silk industry of Kashmir, an opportunity of expressing their views on the present position of the industry for the information of Mr. Dane, who was proceeding shortly to India to take up the post of Resident in Kashmir, previously held by Sir Adelbert Talbot. Mr. Dane had seen the official papers on the subject, and so they would under-

* See the India Office Official Report.

stand that he was aware of the facts connected with the question.

Sir Thomas Wardle said he was strongly of opinion that until the standard of the silk produced in Kashmir was raised to that of the finest Italian silk, and its exact position in the market was definitely determined, the revived industry should continue to be administered by the Kashmir Durbar. There is almost an immediate possibility of the silk being reeled up to the highest European standards by an improved mode of procedure in treating the cocoons which he had suggested for the present harvest, and which had given the silk a better colour and bone. The reeling also was now so much improved that there was no reason why the silk should not soon command as good a price in the European markets as the raw-silk of the Cevennes or Italian silk.

He showed a specimen of the raw-silk, which he had recently received from Kashmir, of this year's crop, which was of excellent appearance and quality. He had sent it to Lyons to be examined, and he read the report, to the effect that, although "duveteuse" (i. s. downy) it was of excellent colour, strength, elasticity, and improved in reeling.

He had been completely surprised and greatly distressed to learn from Mr. Walton that it was in contemplation to transfer the industry to the control of a private company.

The great need now was to afford Mr. Walton extra European supervising assistance, both in the breeding of the worms and in the reeling of the silk; and it was essential for the present that all the profits accruing from the silk turned out by the Durbar should be applied to the extension of the present industry, or at least to the direct benefit of the Kashmir State.

One speaker said that the point on which he was most anxious was the rearing of the silk-worms, quite a separate matter from silk making. In a cottage industry the ordinary amount of seed given to each rearer was one ounce. Now Mr. Walton was getting out 25,000 ounces a season, or enough for 25,000 cottages, when the number of persons employed under Mr. Walton did not exceed 10,000, and was probably nearer 7,000. A large percentage of the cost of the seed was in this way lost; if only from the worms smothering one another when hatched.

In his opinion the proper thing to do was to import "cellular seed" each year sufficient to produce a crop of "industrial seed" for the year following, and to set aside a plot of land for this purpose only, and place the operation under a specialist attending to it only, and to the instruction of intelligent natives in the conduct of it. You would thus gradually educate a number of Kashmiris in the proper methods of rearing of silkworms. If this "cellular seed" proved in any year unsuccessful, there would still be time to get out "industrial seed" from Europe for the next crop. But even if this (impoverished) "cellular silk" had to be spun, the cost of it would be well spent in the education of the rearers.

"Cellular seed" was guaranteed perfectly free from pebrine. But "industrial seed" (*i.e.* seed of the following generation) contains 5, or even more, per cent. of pebrine, or just enough to carry the crop through. The breeders of "industrial seed" do not undertake to give pure seed as some have thought. If they did they would ruin their own business, as the purchasers would then keep enough of their crop to breed again next year, instead of buying from the breeders. As soon as the eggs of the "cellular seed" were laid, the moths could be tested; and if they did not show more than 5 per cent. of pebrine, they would be safe to use for the coming year. If they showed over this, then "industrial seed" should be bought from the European sellers. A special place must be set aside for working the "cellular" seed, as the whole Kashmir Valley was just now full of pebrinous disease, and very contagious. But with so many subsidiary valleys sloping down into the main one, this could easily be managed.

Also the question of working all the year round should be taken in hand at once. If the houses could be kept at a fixed temperature the natives would only be too glad to work in them in the cold weather, as the people do in North Italy, which is much colder than Kashmir.

Sir Thomas Wardle said that the reeling should be continuous, and this could easily be managed by the use of the Sturtevant apparatus, which would maintain a proper temperature in the reeling factories.

He had from the first recommended that the industry should be carried on exclusively with the seed purchased regularly every year from the best races of France and Italy, with the exception of experimental work of a limited nature. The success of the present system in Kashmir from the commencement had conclusively proved the wisdom of this, and in these views he has been long supported by the Government establishments of Padua and Milan, as well as by the opinions of experts and practical sericiculturalists in both France and Italy.

The time had not yet arrived for cultivating silk commercially from eggs raised in Kashmir; he had recommended Mr. Walton to try this at first, but on a small scale only, and it was somewhat curious to learn that each experiment had failed in consequence of the prevalence of pebrine in Kashmir.

The progress under the present system in Kashmir had been marvellous, and pointed to considerable profit, both in the immediate past, the present, and the future, and he most strongly recommended that the present system be continued.

The breeding of silkworms was a very scientific business, and it was difficult to conceive that under the ablest supervision the natives could be successfully taught, in a large way, to prevent disease. He was quite against any system of silkworm breeding other than from purchased eggs, excepting only very gradually. The eggs that he had assisted in purchasing in Italy and France for Kashmir had been absolutely guaranteed to him to be free from pebrine, and he held written guarantees from the dealers. Mr. Walton had written to him very favourably of the freedom from disease of the imported eggs, and he therefore advised them, for the present, to let well alone.

Another speaker stated that the silk was still defective, but in the circumstances marvellous progress had been made; and the information that the Kashmir Government were thinking of making over the new industry to a private company, had come to him as a great and painful surprise. The time had not yet come for the enterprise to be handed over to, or, indeed, taken over by, a private company, and whenever it was passed into private hands it should be under arrangements which, while relieving the Government of all detailed control, gave them a fair share in the profits; and these arrangements should be subject to revision every three or five years. The concessionaires should have a knowledge of both silk production and silk business. If the industry was handed over to mere company mongers, it would be ruined, and the Government greatly discredited.

He would also point out that there is beside the waste a serious risk of importing seed as at present. He understood that the Accountant-General had already pointed this out, and he was quite right; for, if through any cause the shipment were lost, or got chilled on the way out, the year would be a blank, as there would be no time to repeat the shipment before hibernation had set in on this side the water; and once that had begun, the seed would either hatch out before reaching the cold in Kashmir, or get spoilt before it arrived there from the hiatus in the incubation. A typewritten memorandum by two of the gentlemen present on the subject of private enterprise versus State conducted industry was handed in and is hereto attached.

Some general conversation then followed on the subjects brought forward by Sir Thomas Wardle and the others; after which the meeting broke up.

GEORGE BIRDWOOD.

26th September, 1901.

---

1. We beg to submit the following for con-

sideration in connection with the proposal for making over the Kashmir silk industry to private enterprise.

2. In our opinion it is advisable that the State should retain a direct interest in the business for the following reasons:—

(*a*) It is not yet known to what extent the industry is capable of being developed, and, accordingly, it is not possible to fix a fair sum for the State to either exact or accept for the sale of the monopoly. A royalty or tax on silk produced might be found afterwards either unduly heavy or unduly light. In the one case the growth of the business would be hampered and in the other the State would not receive a fair return for the advantages granted.

(*b*) If a great monopoly should grow up in which the State is not directly interested, the interests of the State and those of the monopoly might conflict, and friction arise.

3. On the other hand, there are the following disadvantages in the State itself conducting the business:—

A Government is never well placed for running a commercial enterprise, for its officials cannot have the requisite freedom of action; indeed, it would be hardly warrantable for them to act in the same way as mercantile men with experience of their trade, and with their own money at risk. With Government working, the control of expenditure must rest with the finance authorities, who must frequently find difficulty in determining whether the recommendations of the silk department should be adopted or not. In the present case we gather that the following difficulties have arisen, and these and others can hardly fail to occur from time to time.

(*a*) Money is difficult to be had when suddenly wanted, as may often be the case in a business of this sort.

(*b*) Sanction for additional assistants is withheld though necessary for proper reeling, and thereby the value of the silk may be lessened by 10 to 15 per cent,

(*c*) Estimates are cut down and a large increase of good business is thereby likely to be lost.

(*d*) Estimates on an excessive scale may be passed. At present looking to the quantity of seed bought, and the silk produced, it is evident that a large portion of the seed is lost, which shows extremely bad education, and a great expense could be avoided if the natives were better instructed in the work.

In our opinion the best course is that the State should retain a direct interest in the business, but that it should rid itself of all control and management of it, and hand it over on payment of an agreed sum to be conducted as a private enterprise, the profits to be divided in proportions to be arranged, between the State and the concessionaries.

This course would best secure the good management and expansion to the utmost of the industry. It would free the State from the difficulties and risks of running the business on its own account, for it should be provided that the State cannot be called upon to pay any losses. Moreover, in our opinion the profits which the State would receive would in all probability much exceed any profits it would make under the present system.

5. The power of recalling the concession after a fixed number of years, and on fair terms, should remain with the Government.

6. The concessionaires should have a knowledge of both silk production and the silk business. If the concession were handed over to company promoters the result would very probably be the ruin of the industry.

CHAPTER XII.

A CONFERENCE (THE THIRD) ON SERICICULTURE HELD AT THE RESIDENCY, SRINAGAR, ON FRIDAY, MAY 1ST, 1903, IN THE PRESENCE OF THE RESIDENT IN KASHMIR, DURING THE VISIT OF SIR THOMAS WARDLE.

PRESENT :* The Governor of Kashmir; Major J. L. Kaye, Settlement Commissioner; Mr. F. Anderson, Accountant-General; Mr. Walton, Director of Sericiculture. Sir Thomas Wardle also attended the meeting, by special invitation.

Read proceedings of the last conference of 8th September, 1902.

The following matters came under discussion.

Seed.

1. † 1-2. Seed has been sent to the Ramband Tahsil for this year. Sir T. Wardle is opposed to the idea of raising industrial seed in the country, but thinks that a small experiment may continue to be made with portion of this seed.

* The Revenue Member of Council was unable to be present, but has since signified his assent to all the decisions arrived at in this Conference.

† Reference to previous meetings.

Decided that a portion of the seed produced should next year be sent, after microscopical examination, to another part of Kashmir to be reared.

2. *I-6. Mr. Walton has sent seed to the Tahsildar of Ramband and Kishtwar, and about 100 rearers are being employed in these two tahsils this year.

Mr. Walton will shortly get full particulars from the Governor of Jammu.

A few rearers in Bhadrawah have also been given seed.

Decided that Mr. Walton should report progress and results after six weeks.

3. *II-1. Seed has been distributed on the principles approved, to a largely increased number of rearers. The total number of rearers is likely to be fully 10,000. The Committee expressed approval.

4. Decided that 24 maunds be obtained for next year instead of 20 maunds, and that the Director get half a maund of Italian seed from Italy to experiment with.

Rearing.

5. *II-3. In this para. 20,00,000 strong young plants should be 2,00,000.

The number has now been raised to 3,15,000, exclusive of Verinag.

The Verinag nursery, however, is found to be

* Reference to previous meetings.

damp, and may have to be abandoned. The same number of trees will be planted in one of the other places. Mr. Peychaud will be asked to continue the process every year.

Sir T. Wardle's severe criticisms of the manner in which the mulberry trees had been damaged by the loppings for feed last year, and by the unscientific pruning this year were considered. As regards the latter, the Resident stated that he had enquired into the matter and had ascertained that the Governor of Kashmir, in spite of several letters from the Director of Horticulture, had delayed the issue of orders until much too late in the season, and had then issued urgent orders, which led to the pruning being carried out in a hurried and careless manner. The damage that was being done had, however, been discovered before it had gone very far, and was promptly stopped. It was agreed that the so called pruning had chiefly taken place along the river and roads, where it was most noticeable, and had not had time to extend to villages at a distance, before it was stopped. This opinion has been verified by further enquiry.

In regard to loppings for feed, it was stated that the villagers were allowed to cut branches much too large, but the complete prohibition of cutting as advised by Sir Thomas Wardle would be dangerous to carry out in Kashmir, where the people have no ladders or rearing nets, as in

France, and that the worms might not get sufficient air and ventilation if leaves only were to be picked.

Decided that for this season lopping for feed should be restricted to shoots not thicker than a man's finger.

As to pruning that this should ordinarily be done in Sawan as already directed; that Mr. Peychaud should be asked to tour at that time this year, to instruct the villagers in the right method of pruning, in all Tahsils if possible; that for the future an Assistant to the Director of Sericiculture should be appointed who should be a qualified Forest Officer, on the ordinary scale of pay for Sericiculture Assistants, with fixed travelling allowance. His duties will be to supervise the Inspectors of rearing, of whom there are now three in each Tahsil, to prevent undue lopping of trees for feed, to instruct the people in pruning, to see that their rearing houses are kept clean and in good order, and to give them general advice as to methods of rearing.

The committee would also place under this officer and the inspectors the village nurseries, which are not, under present arrangements, working well. The Tahsildar would be required to see that any matter which the Inspectors could not themselves put right were duly carried out by the Lambardar.

Sir Thomas Wardle suggested that the attention of the officer should be specially directed to the cultivation of the white mulberry and the "bedana." The Committee accepted this suggestion. Mr. Peychaud has reported that great numbers of the "white mulberry" are already in his nurseries.

6. The question of rewards for rearing was considered.

Decided that rewards should not be given to Tahsildars and Lambardars, but that prizes should be given to rearers, who have their silkworms and arrangements in best order, and to Lambardars of villages which, on the whole, show most care in treatment of trees and in care of the worms.

Weighment and Storage.

7. * III-1. Accountant-General reports that he has considered the opium rules, but is of opinion that opium being a very valuable commodity in small bulk, the conditions are very different and the rules are much too elaborate to apply.

The Committee consider that the best system is, that the cocoons as brought in, should all be weighed out by assistants, whose accounts should be recorded by themselves, and by a clerk, in the vernacular, the results being entered at once in the tally book, and thence transferred to the Filature Account Books. The weighments last for twenty days, and the present arrangements for supervision are regarded as desirable. There are

* Reference to previous meetings.

three scales, and if, owing to the new filatures, additional assistants are necessary, the inspection work in the filatures can be entrusted for this short period to the assistants not employed on weighments.

8. * III. (2) (First meeting). Decided that the payments by results is sufficiently secured by the varying amounts produced by different rearers from the same amount of seed, though men who bring in inferior cocoons are paid at Rs. 10 per maund, or some figure lower than the average rate, and these men are further punished by getting less seed next year.

9. A question raised by the Resident, of keeping the cocoons of different Tahsils separately, in order to test by results the silk produced in different localities, was considered.

Decided that a sufficient quantity of cocoons from each Tahsil should be kept separate and treated separately throughout, for experimental purposes: and that the results in value of silk produced from each, should be duly reported by the Director.

10. * III-4. One séchoir has been provided by the Public Works, but the cost of this is still not reported. State Engineer to be asked to expedite this. Seven séchoirs at Rs. 2,500 each, have been ordered from Hari Ram & Co., of Rawal Pindi; one has arrived and two are on

* Reference to previous meetings.

the way, and four are to leave on the 15th May. The boilers and engines and fans left Pindi on 25th April. The State Engineer has been requested to expedite the erection of the séchoirs and boilers in place.

Director to be asked to report on the Hari Ram séchoirs after they have been used this season.

Reeling and Filatures.

11. *IV. 2, 3. The additional Filatures for which at the last meeting it was decided to make provision, are approaching completion, and are single storied buildings with excellent ventilation. The engine houses are not quite complete, but the State Engineer has been pressed to push these on, and also to put the boilers, which have arrived, in place. It is anticipated that the four filatures will be all ready, and in working order, in June.

Decided that immediate steps should be taken for the insurance of these new buildings.

12. *IV. 1 & 2. Various proposals for economy of working were next discussed. The Director reported that the experiment of putting up a verandah for fifty basins had been tried, but that the effect was to darken the interior of the building, and make it more difficult for the reeler to see flaws in his silk there, while in the verandah itself the workmen had to sit with their

* **Reference to previous meetings.**

faces towards the light, whereas for proper detection of duvet, &c., the light should be behind. The State Engineer had also reported that the construction of the verandahs, considering the smaller number of basins for which room could be provided, would be scarcely less expensive than the construction of new filatures.

The Committee decided that the verandah experiment was not likely to prove economical, but that the Director should submit a special report on the value obtained of the silk wound in the verandah, and in the filature immediately opposite the verandah.

In regard to electric lighting, expert opinion is to the effect that it is bad for the quality of the silk, a view which is fully endorsed by Sir Thomas Wardle. The Resident pointed out that the use of electricity would, however, be in many ways advantageous *e. g.* for power-reeling, a form of econony which is strongly advocated by Sir Thomas Wardle. The Accountant-General pointed out that electric heating should also be considered. At present the consumption of fuel is enormous, and it would in time, as the industry grew, constitute a great danger to the forests of Kashmir, or to the industry itself, and that this danger would be greatly reduced, if not altogether avoided, by the application of electricity for heating.

Decided that the first experiment towards economy of working be directed towards power-reeling, and the application of electricity both for reeling and for heating, and that State Engineer be asked to consider the question of providing water-power for the purpose, and further that Messrs. Osler & Co., Kilburn & Co., and Messrs. Octavius Steel be addressed by him for full information with regard to methods of electric heating.

13. *V. 2. The Resident pointed out that owing to the heavier snowfall this winter, there would probably be no difficulty this year in floating timber down the Dudhganga, and that the Conservator of Forests had informed him that a large supply of fuel was to be brought in by this river. The State Engineer would, however, consider the question of canalisation of the Dudhganga together with the subject of providing water-power near the filature, and probably both schemes would be worked together, though electric heating could be arranged, the necessity for providing water-carriage for fuel for the filatures would be less imperative.

Establishment.

14. *VI. Decided to urge Revenue Member to nominate sericiculture apprentices without further delay, and that the number ought to be at least five.

Meanwhile the necessity of further European

* Reference to previous meetings.

supervision to supervise work in the additional filatures is incontestable. Sir T. Wardle urged the view taken in his Report that there should be one European assistant to each filature building, but the Resident pointed out that in view of the eventual appointment of high class Native supervision, so large an increase was not desirable. The Director reported that Mr. Phillips had resigned his appointment.

Decided that two more European assistants should be engaged, one to replace Mr. Phillips and one additional, thus securing one such assistant to every two filatures.

Mr. Phillips having resigned, at present there are only three not four.

15. Sir T. Wardle's proposal about sending young men to France to be trained was considered. The Resident pointed out that this would secure men with previous training, but could not be carried into effect this season, whereas the question of appointing additional assistants was immediate.

Decided to accept proposal in regard to one such assistant, and that of the two to be now engaged in India, one should be engaged permanently and the other temporarily.

16. The Settlement Commissioner raised the question of house accommodation.

Decided that a house with accommodation for four assistants should be built on the site originally selected for that purpose in time of Sir A. Talbot.

Sale.

17. *VII. The Committee agree unanimously that the silk should continue to be trusted to one agent only, and that all business opinion is against giving it to more than one agent in Europe. Messrs. Durant Bevan have done very well, and the Committee decide not to divide the produce but to continue to entrust it to them, as at present, as sole agents in Europe.

When the silk is sufficiently good for the American market, the question of employing a separate agency for America can be considered. Sir T. Wardle can advise as to this.

18. Mr. Walton laid before the Committee a letter from Messrs. Durant & Co. with a complaint about certain of the silk supplied being moth-eaten. This complaint appears to be only from one buyer, and applies only to bales bought by him. No complaint has been received about other bales sent at the same time.

Decided to wait another letter from Mr. F. Durant who is going to call at Marseilles, on his way to England, to look into the matter.

This is the first objection ever received of the kind. The bales are packed in waxcloth and gunny. The bales complained of were dispatched

* Reference to previous meetings.

at different times, during a period extending over four months, and no other purchaser has complained.

Accounts.

19. *VIII.-i. Mr. Anderson thinks that the account forms are sufficient, and might indeed be simplified.

The stock of silk has been checked. The forms have been in use for six months, and have been found to be sufficient with some simplification.

The Accountant-General proposes stock should be taken twice a year, at the end of the official year and in June, in order to check at a time when the cocoon stock is likely to be low.

Proposals Approved.

20. *VIII.-ii. Rules for regulating the power and position of the Director have been passed and approved.

21. The Accountant-General laid on the table a statement of account (copy attached) showing that since Samvat 1954, the business now showed a profit of Rs. 6,03,308-10-3, inclusive of part of commission due, or a return in six years of 117 per cent. on the capital expenditure. During the present year the actual cash receipts, exclusive of stock and sums due but not yet credited in the Treasury, had for the first time, exceeded the total working expenses. The cocoons in stock had

* **Reference to previous meetings.**

been valued at Rs. 15 per maund only, whereas their value when converted into silk, is likely to be much higher, so that the estimate of profits is, if anything, below the mark.

The account was explained by the Accountant-General and accepted.

22. The Accountant-General laid before the meeting a proposal that the value of silk should be realised from the agents at once, instead of after three months interval. The State would, by this, lose the discount, but as the rate of interest in India is generally higher than the bank rate in England, if the money were put out to interest immediately on arrival in India, there would be no loss, and the State would avoid any risk which might arise from the money being left unsecured in the hands of the agents during the three months' interval.

Decided that the agents should be asked to realise at once, the Accountant-General being responsible for putting the money received out to interest at once.

Miscellaneous.

23. Employment of female labour.

Decided that the reelers and turners should be informed that if they wish to bring their women-kind to work with them, thereby adding to the family earnings, they would be permitted to do so. Arrangements would be made for the male and female members of one family to work together

24. New filatures.

Decided that the State Engineer be asked to select a site for a filature at Anantnag, and to report within four months whether it could be worked with electricity, and to prepare estimates of cost, both for a filature worked by steam, and one worked by electricity. The estimate might be for a filature of 300 basins.

25. Vote of thanks to Sir T. Wardle.

The Resident said that Sir T. Wardle, ever since 1896, had taken the keenest interest in the industry, and had offered the most disinterested advice. He had now come out to Kashmir at his own expense, and had written two reports which were with members (copies attached) in which he had embodied his views on the whole subject. Most of Sir T. Wardle's suggestions had been accepted by the Committee. The Resident thought that the industry was much indebted to Sir T. Wardle for the help and counsel he had given for so many years.

Resolved that the Resident's remarks be endorsed and a vote of thanks recorded to Sir T. Wardle.

Extract from an Un-Official Note dated the 1st May 1903, from the Accountant-General, Kashmir State, to the Resident in Kashmir.

Preliminary Balance-Sheet of Silk Industry from

Samvat, 1954 (1898), to end of 1959 (12th April, 1903).

LIABILITIES.

		In Rupees.			In English money.		
		Rps.	A.	P.	£	s.	d.
A.	Capital	5,16,320	9	1	34,421	6	8
B.	Working expenses	17,07,895	8	9	113,859	14	8
C.	Profit	6,03,308	10	3	40,240	8	0
		Rps. 28,27,524	12	1	£188,521	9	4

ASSETS.

		Rps.	A.	P.	£	s.	d.
D.	Block	4,10,935	12	10	27,395	13	4
E.	Cash receipts ...	17,26,261	15	3	115,084	1	4
F.	Stock in hand ...	6,90,327	0	0	46,021	16	0
		*Rps. 28,27,524	12	1	£188,501	10	8

Out of profits, Rs. 3,476-6-0 have been disbursed as part of the commission due on profit of earlier year's working. A further sum is now due.

When all profits in connection with the accounts for 1959 are settled, a revised balance-sheet will be issued.

CHAPTER XIII.

PROCEEDINGS OF THE FOURTH CONFERENCE ON SERICICULTURE, HELD AT THE RESIDENCY, SRINAGAR, ON THURSDAY, OCTOBER 22ND, 1903, IN THE PRESENCE OF THE RESIDENT OF KASHMIR.

PRESENT: General Raja Sir Amar Singh, K.C.S.I.; K. B. Ghulam Ahmed Khan, Revenue Member, Kashmir State Council; Major J. L. Kaye, Settlement Commissioner; Pandit Man Moham Nath Kaul, Governor of Kashmir; Mr. G. M. R. Field, State Engineer; Mr. F. Anderson, Accountant-General; Mr. C. B. Walton, Director of Sericiculture; Major A. F. Bruce, Assistant Resident.

Read proceedings of the last Conference held on the 1st May, 1903.

The following matters came under consideration.

Seed.

1.—1.* Mr. Walton reported that the cocoons raised in the Ramban Tahsil had been subjected to microscopic examination, and the seed was found to be tainted with disease and unfit for

* Reference to Proceedings of 3rd Conference.

distribution as "Industrial seed" in any portion of Kashmir as was decided at the last Conference.

Decided, that the danger of using tainted seed could not be overestimated and that the cocoons should be used for reeling and not for seed.

2.—3.* The actual number of rearers this year was 10,898, as against 8,158 in 1902, and the average amount of seed per rearer 2·34 oz. as against 3·12 oz. in the preceding year. Owing to unpropitious climatic conditions the seed had to be distributed later than usual and the crop suffered from excessive heat, as well as from abnormal cold following rain. The total crop of the year was more than 25 per cent. less than in 1902, the decrease being due to the above-mentioned causes, in spite of the larger number of rearers among whom the seed was distributed.

The Resident has brought to notice, that the amount of seed given to some rearers is still too large. These men again distributed the amount they are unable to rear themselves to "kashtkars" and take a portion of the crop from them. This is objectionable. No one man should be given a larger amount of seed than he can rear himself.

The Settlement Commissioner also brought to notice, that when distributing seed, sufficient attention was not paid to the supply of mulberry leaf available in each village. Sometimes a

* Reference to Proceedings of 3rd Conference.

village possessing few mulberry trees was given an amount of seed for which feed in the village was not available and that in neighbouring villages the amount of feed was only sufficient for the seed. This leads to continual disputes among the rearers, and as the total amount of feed available in the neighbourhood is insufficient for the seed distributed, the crop consequently suffers.

Decided that, especially in view of the larger amount of seed ordered for next year, every attempt must be made to still further increase the number of rearers and that now that the industry has gained popularity, the Director must give special attention to the capacity of each individual rearer and to the amount of mulberry leaf available in each village when deciding the amount of seed to be distributed to the rearers.

3—4.* The extra amount of seed decided on at the last Conference has been purchased and has arrived.

4. The Resident noticed that there were a a considerable number of mulberry trees in the Naoshera Tahsil Jammu and that operations might well be extended to that tahsil.

The Director stated that he was in communication with the Governor of Jammu on the subject.

It is noted that the seed will probably have

* Reference to Proceedings of 3rd Conference.

to be distributed earlier in this tahsil and the Director should determine the proper time in communication with the Governor of Jammu.

Rearing.

5.—5.* Owing to the recent floods the Director of Horticulture was unable to proceed to each tahsil during the month of Sawan to superintend and give lessons in pruning. He has promised to visit each tahsil during the winter months and direct pruning operations where pruning may be found necessary.

Decided that the Director of Horticulture should tour in the winter months as promised. The Committee would note, however, that, as trees pollarded during the winter will be useless for the supply of leaf during the ensuing spring, pollarding should only be effected to a limited extent, and in villages where the number of mulberry trees is very large.

6.—5.* The Settlement Commissioner brought to notice that the damage done to mulberry trees by rearers felling large boughs is still very great. Practically no attention was paid to the orders directing that no boughs thicker than a man's finger should be cut. The Inspectors of rearing do not seem to perform their duty in preventing such damage. When the damage done to the trees has been excessive, punishments have been

* Reference to Proceedings of 3rd Conference.

given this year and it is hoped that the effect of such punishments may be apparent in more careful lopping for feed next year.

Recorded.

7. The Revenue Member having brought to notice that the large amount of mulberry leaf available in the Dachigam Valley was not available for use by the rearers, correspondence on the subject has been conducted with the responsible officers. Colonel Ward, as officer-in-charge of the State Shikargahs, has agreed to allow leaf to be taken from the Dachigam Valley and is in correspondence with the Director on the subject.

Decided that the Director of Sericiculture should notify to the Vice-President the dates during which mulberry leaf is required from Dachigam.

8.—5.* In accordance with the recommendations made at the last Conference, the State Council have sanctioned the entertainment of a forest trained assistant.

The Director has heard from the Superintendent of the Dehra Forest School that a suitable trained man may be available next March and is in communication with the Conservator General of Forests on the subject of the entertainment.

The action so far taken is approved, and it is hoped that the matter will be settled before the close of the winter.

9.—6.* The Director of Sericiculture expressed

* Reference to Proceedings of 3rd Conference.

his inability to frame any suitable proposals for the grant of rewards to rearers and lambardars.

The new Forest Assistant when appointed will be asked to give his special attention to this question of rewards.

10. The Resident pointed out that the question of State-built rearing-houses had been left over for further discussion after the results of the present year's crop was known (vide Proceedings of 2nd Conference II-2).

In his opinion large houses, in which a number of men would rear silk-worms, were not required. These are difficult to keep clean and it is doubtful whether the rearers would make use of them. He was inclined to advocate the building of small houses sufficient for one man. These could be left to men whose objection to taking seed is that they have no room for rearing. The rent could be recovered from such rearers' cocoon payments.

Or, as an alternative suggestion, he would suggest that loans might be given by the State for the building of such houses, the sums advanced being recovered in the same way in two or three years.

The Committee are in favour of the second proposal coupled with the condition, that the lambardars are held responsible, that the money is applied to the purpose for which it is advanced and that the amount in each case does not exceed

Rs. 50. The loans given in the first year would be free of interest.

Weighment and Storage.

11.—10. The engines and fans supplied for the new séchoirs by Messrs. Davidson & Son have not worked satisfactorily. Whether the defects are due to faulty installation or to inherent faults in the machinery is being tested. Under the sanction of the State Council the suppliers are sending up a man to examine the machinery.

The séchoirs supplied by Messrs. Hari Ram & Co. have given satisfaction.

Recorded.

Reeling and Filatures.

12.—11.* The four new filatures have been completed since the last Conference and are now in use. Two of these buildings have been set apart for women reelers. The buildings have been insured.

Recorded.

13.—12.* In regard to the adoption of electricity for power-reeling and heating, the State Engineer reported that he would furnish a note on the subject. (Vide Appendix).

The Committee decided that the experiment should be made but that the question of the application of electricity should await the develop-

* Reference to Proceedings of 3rd Conference.

ment of the general water-power scheme now under consideration.

14.—13.* The questions of the canalisation of the Dudhganga, and of providing water-power near the filatures, are being considered in the Public Works Department in connection with the Flood Spill-channel scheme.

The canalisation of the Dudhganga, might be prepared but should not be carried out until the experiment in electric heating has been made.

15. Pending the elaboration of schemes for improving the carriage of fuel to the filatures by water or for the adoption of electric heating, a tramway has been laid down to connect the filatures with the river, this has proved a great convenience and economy in the transport of fuel from the ghat by the La Mandi to the silk factory. The line was damaged by the floods but is now repaired.

Recorded.

16. The experiment in power reeling, the power to be generated by the séchoir engine, has not been tried. The State Engineer, in accordance with the decision arrived at by the Committee at the Second Conference (Proceedings of Second Conference IV. 3.) prepared an estimate for a separate shed to contain 20 basins. The estimate amounted to Rs. 4,073 and this was considered a large sum to pay for the experiment. The

* Reference of Proceedings of 3rd Conference.

question of applying power to 20 basins in one of the present filatures was suggested, as an alternative, and the State Engineer was asked to have a revised estimate prepared on those lines. The estimate has not yet been received.

Decided, that the State Engineer will frame a new scheme for the experiment in power reeling, making use of existing buildings and machinery as far as possible.

17. The Committee notice with pleasure that continued improvement is taking place in the quality of the reeling, as evidenced by the prices obtained for the silk. In 1901 the average price per lb. was only 13s. 3d. In 1902 the better silk realised 15s. per lb. (vide Proceeding of Second Conference IV., 1.) The last sales reported this year show that the price obtained has now reached 16s. 11d. per lb. This high value proves that the Durbar is getting an excellent return for the money expended on extra supervision.

The Committee whilst noting with satisfaction the high prices realised suggests that Messrs. Durant Bevan & Co. be asked to quote the prices realised by silk of the same denier from Italian and French filatures in the same years.

Establishment.

13.—14.* In accordance with the decision of the last Conference, the Revenue Member on the

* Reference to Proceedings of 3rd Conference.

10th June last selected five candidates for training in sericiculture work.

1. Thakar Das Bali.
2. Pundit Aftab Kaul.
3. Pundit Shiv Ram.
4. Pundit Maheshor Kaul.
5. Khawaja Ghulam Mohi-ud-din.

These men have now been in training for four months. The Director reports that these are absolutely useless and recommends their dismissal. He states that they are not men of the right class, have not the educational qualifications laid down by the Second Conference (Proceedings of Second Conference VI., 1,) and do more harm than good, being unable to command and only give incorrect orders.

He advocates the promotion of the Head Filature Inspectors to the post of native assistants, as these men are trained, will see orders carried out and are of just as good family as the candidates deputed while such promotion will encourage good work.

Decided that the candidate who has given least satisfaction should be dismissed but that the Director should give the others a further trial punishing them when necessary in order to enforce discipline.

19.—15.* Sir Thomas Wardle intimates that he has selected in Europe a Mr. Thomas as an

* Reference to Proceedings of 3rd Conference.

assistant and sent him to France for training. His training will be completed and he will be available for appointment about March, 1904. Sir Thomas has enquired whether the State will defray the cost of his passage from England to Kashmir.

The Committee advise that the cost of a second class passage be paid by the State.

20.—14.* According to the advice tendered by the Committee at the last Conference, the State Council sanctioned the entertainment of one more permanent European assistant in place of Mr. Philips, resigned, and the engagement of a second such assistant as a temporary measure, pending the arrival of the assistant selected by Sir Thomas Wardle for training in France.

Mr. Walton reported that he was unable to obtain the services of a European assistant on the terms offered for temporary employment only, and that as extra assistance was necessary he engaged both Messrs. Amesbury and Giles for permanent employment, subject to their giving satisfaction while on probation.

Mr. Walton asks that his action be approved, and that, in view of the increased work caused by the opening of the new filatures and the extra testing of silk, both assistants be permanently employed.

Decided that the position as already arranged

* Reference to Proceedings of 3rd Conference.

by the Committee be carefully explained by the Director to Messrs. Amesbury and Giles as the services of one of them, unless a vacancy occurs in the meantime, will probably have to be dispensed with. The temporary appointment of one assistant as sanctioned at the third Conference was intended to be for one year.

21. The sericiculture assistants have sent in a memorial setting forth that, in consideration of the long hours of work taken from them, the present rate of salary offered by the Durbar (Rs. 150- 25- 300-) is too small and asking for more generous treatment.

Decided that no sufficient grounds have been disclosed for an increase of salary.

22. The calculation made by the Accountant-General of the amount of commission due to the Director up to the end of Samvat 1959 was laid before the meeting.

Decided that the Accountant-General be asked to re-examine the figures for the first two years 1954-55 with a view to ascertaining the exact loss which has accrued to the State during those two years and which should not be taken into account in calculating the Director's profit. Subject to this proviso the Director should then be called upon to elect whether he accepts Mr. Anderson's figures as they stand, or the method of calculation adopted by Mr. Pritchard.

Sale.

23.—18. With regard to the complaint of certain silk having been received moth-eaten, enquiry has shown that the damage caused was by friction in carriage and not by moths.

Although every endeavour is made to pack the bales in such a manner that the silk may not be rubbed, it is very difficult to ensure that such damage is not caused during the long transit to Europe. The portion of the journey most to be feared is that between Srinagar and Rawal Pindi by cart. The Director reports that the carts supplied by the Imperial Carrying Company for the purpose are most unsuitable, being roughly made with projecting nails, etc. To safeguard the silk on this part of the journey the Director has been in communication with Messrs. Hari Ram & Co. of Rawal Pindi, who, on condition that all silk and stores required by the Sericiculture Department are made over to them for carriage, agree to build special carts for the carriage of silk and to charge Rs. $\frac{3}{8}$ per maund for transport to Pindi instead of Rs. 4 charged by the Imperial Carrying Company.

The Director has been informed that he *may* make a contract on these lines.

Recorded.

24.—17. The improved value obtained for silk this year has been noticed in paragraph 17 supra. The Committee record their appreciation of the services of the Agents for the sale of silk, Messrs. Durant Bevan & Co., in this connection.

Miscellaneous.

25.—23.* The introduction of the scheme for the employment of female labour in the factory has been effected and appears likely to prove a success. One and a half filatures are now filled with female reelers, and though these are of course still backward, some of the women are now reeling four skeins.

The Committee express their satisfaction at the results obtained in this direction, and note that, if the scheme continues to be attended with success, the difficulty of the permanent supply of labour and the retention of skilled hands will be overcome.

Decided that the filatures be kept open as late as possible this year in view of the scarcity prevailing in the city.

26—24.* The preparation of plan and estimate for a filature at Anantnag, to be worked either by steam power or electricty decided on at the last Conference, has not yet been carried out. The question of the site, which was left to the Public Works Department, has been referred for consideration by the Sericiculture Department but no further communication received.

The Committee note that, in view of the shortage in the crop of this year, no such addi-

* Reference to Proceedings of 3rd Conference.

tional filature will be required at present. For future consideration the plans and estimates should however, be prepared and the State Engineer has been asked to arrange for this.

27. The supply of fibre for the brushes used by the cooks is difficult to procure. The old system of indenting on village labour to supply the fibre is unsatisfactory. The Director states that he has been unable to obtain a contractor for the supply of fibre.

The Settlement Commissioner states that he has asked the Governor to try and find one or more men willing to contract for this supply and to put the Director in communication with them.

Recorded.

28. The Director reports that the old Sericiculture buildings at Chirpur and Raghnathpura in charge of the Sericiculture Department are no longer required and might be dismantled.

The Settlement Commissioner states that he has addressed the Revenue Member with a view to the buildings in question being taken over by the Revenue Department or dismantled as suggested.

The Committee agree that the buildings should be dismantled and the chaukidars brought under reduction.

29. The question of the introduction of silk-weaving in Kashmir advocated by Sir T. Wardle was discussed.

Decided that enquiry be made frcm leading

firms in Bombay as to the value of the silk after it has been woven as compared with the raw article. If the results of the inquiry are satisfactory a dozen looms might be started at some central place near the filatures and Mr. Stott engaged on the terms proposed.

Appendix to the proceedings of the Fourth Conference on Sericiculture.

Note by G. M. R. Field, State Engineer, on the question of heating the water in the basins by electricity in the Silk Filatures, Srinagar.

1. Some little time ago the Resident in Kashmir asked me to make enquiries regarding the heating of water in the basins at the silk filature by electricity I took up the matter but was quite unable to find anything about it in the Indian catalogues. I therefore wrote to several of the largest firms in Calcutta and received some replies. But they were all very unsatisfactory. No one seemed to know anything about such a process nor could they give me any information. Several firms asked for more particulars and these were supplied but up to date no information has reached me.

On the arrival of Mr. Simkin, the Electrical Engineer, I discussed the matter with him. He told me that as far as he was aware very little had been done in this direction in England or on the Continent. That the apparatus was very expensive and uncertain in its application. The breakdowns were frequent and restorations difficult and expensive and in fact almost prohibitive in a place like Kashmir so far removed from any large manufacturing towns, etc.

3. There are two obvious methods by which the water can be heated.

(1.) By heating water in the boilers as at present only using electricity instead of fuel.

(2.) By heating water in the basins by means of electric currents passing through the filatures on wires.

The great objection to the first method is the uncertainty of operation. If a breakdown occurred the whole filature served by the boiler would be at once thrown out of work. The repairs would be very costly and probably delay the re-opening of the filature for weeks or months.

By the second method if a basin broke down no great harm would be done, it could easily be replaced by a new basin (of which a reserve stock could be kept) at trifling expense.

4. On the whole, therefore, the second method was proposed as the most efficient. Mr. Simkin then wrote off at once to Messrs. Crompton & Co. in England as being the most likely people to be able to satisfy our demands. Mr. Simkin sent a sketch with dimensions of the existing basins and gave full particulars of the work to be done, the temperatures required and other details. It was

suggested that each basin should be separately heated and the current controlled by a regulating switch. The resistance coils in the heaters should be arranged so that they could easily be replaced if injured, and should be as simple as possible. Messrs. Crompton & Co. were also asked if they would supply six such basins at a very early date for experimental purposes. They were also to quote approximate cost and power required. They were also asked whether it was possible to arrange an automatic switch which would cut off current at a certain temperature. To the above enquiries no replies have as yet been received though expected every mail.

5. In addition to the above, the Resident in Kashmir has made enquiries as to reeling the silk by power instead of by hand. This matter was talked over with the Director of Sericiculture and he said that he did not think it would work on account of the practical difficulties of controlling and regulating the speed, etc., while the reelers were at the same time handling the cocoons and threads. However I discussed this matter also with Mr. Simkin who considered that when he got our basins from Messrs. Crompton and Co., the best plan would be to put a rough shed at the electric installation in the Basant Bagh and set up our basins and the reeling wheels in connection with the electric light engines. We could then at the same time test

experimentally the heating of water in the basins and the reeling by electric power. This suggestion seemed to me a very good one as it would save the heavy expense of the wires out to the silk factory. All that we should require would be a small motor costing perhaps Rs. 600 and some shafting and a few pulleys. The expense would be trifling and the experiment likely to be satisfactory alongside the existing engines and dynamos.

CHAPTER XIV.

MY NOTES ON THE PRESENT STATE AND METHODS OF SERICICULTURE AND COGNATE SUBJECTS MADE DURING A VISIT TO LYONS AND IN THE CEVENNES, CHIEFLY AT ALAIS (GARDE), FRANCE, IN FEBRUARY, 1903, EN ROUTE TO KASHMIR, BEING REPORT No. 1, PRINTED IN JAMMU.

On Cellular Seed.

CELLULAR seed or "graine" are the eggs of the silkworm moth Bombyx mori laid in cellules, each cellule enclosing the eggs laid by a female moth, as well as the moth itself.

There are two kinds of cellules (both being small openly woven muslin bags), the smooth cellule made in frames, and the dry or wrinkled cellule. The latter should always be used as they offer every desirable guarantee to the purchaser. One hundred cellules contain one ounce of 30 grammes of eggs, graine or seed, but generally 100 cellules contain 35 to 37 grammes of seed. The former cellule, is called the smooth cellule, the latter the pouched or wrinkled cellule, because the latter is pouched or folded. All seed should be refused except those sold in pouched or wrinkled cellules. This is the kind sent year by year by Messrs. Laurent de l' Arbousset and Son to Kashmir.

The seed sent in this way in pouched cellules or bags gives a guarantee, that the moth in each cellule has been microscopically examined for pebrine, and shows that greater expense has been incurred, because each moth has to be crushed separately and microscopically examined.

The following details explain why smooth cellules cannot be safely guaranteed.

The first case is to make a selection of the best moths so as to keep up the breed and prevent degeneration. An expert employé is appointed whose duty it is to select from those cocoons chosen for reproduction or breeding, the finest couples or moths, male and female, and to put them into cellules which are called cellules of production. When the eggs are laid, M. Arbousset and his son solely undertake the microscopical examination of the moths.

Only those eggs are retained of which the moths have passed a good examination as to breed, freedom from pebrine and other defects.

These eggs are then distributed in Var some 150 miles South-east of Alais in small quantities of about 10 to 25 grammes at the most, amongst 600 or 700 growers for rearing, which has become a cottage industry. These rearings are visited four or five times during the hatching either by

M. Arbousset or his representatives and the progress entered in his books by name of the rearer and number of the order. All the rearings which have shown any signs whatever of disease during the hatching are pointed out and sent to the filature on their arrival at the cocoonery. Rearings which have not shown any signs of disease arrive at one of Messrs. Arbousset's offices. The Manager weighs the cocoons and separates at once one kilogramme. These are counted so as to know the number per kilogramme.

The counting done, the next thing is to shake each cocoon and listen, in order to know how many dead cocoons there are per kilogramme.

The dead cocoons are set aside and taken to the Director of the cocoonery, who opens them one by one to find out what disease the chrysalides have died of, whether of flacherie, grasserie or muscardine.

The number of cocoons per kilogramme is taken down for each rearing and every Manager is ordered to report them to the filature.

Every rearing of the Var race should have more than 450 cocoons per kilogramme.

Any rearing which has not given at least two kilogrammes of cocoons for one gramme of graine hatched is rejected also.

Also every rearing having more than 3 per cent. of deaths through any kind of disease. At the same time a special woman picks out of each rearing the double cocoons (called in Italy doppio cocoons), these are where two silkworms have made one cocoon and in so doing have rendered it unreelable. She opens them and takes the chrysalides to the microscopist to be examined. If the examination is satisfactory, the sorting of the rearings is begun. It is done on a very large table, where they separate all the satiny-looking cocoons, also the saffron or over-coloured cocoons, and all the small and pale ones which have too poor a fibre, as well as all the dead or badly proved cocoons that may be found.

When each rearing is thus prepared and sorted it is ready to be set aside on the cocoonery to wait for the emergence of the first moth and then to place them on the frames or shells.

As soon as the moths appear and it is possible to select 100 couples of them, they are carried into the examination room and put under microscopic examination. If they do not show any signs of disease the rearing is admitted into the cocoonery.

If there should be only 2 per cent. of pebrine corpuscles found, such moths are either sent to the head office or put into the stifling stove attached to the Establishment at Besse.

This is M. Laurent de l' Arbousset's system and knowing for several years his great experience and the high confidence in which he is held both in the Cevennes, in Var and the districts outside France,

Hungary amongst others, where sericiculture has been successfully introduced, in recent years, I feel sure that the best selection possible has been made from the number of French and Italian dealers in silkworm-eggs upon whom I called when sent down to these countries in 1897 by the Government of India.

The result has satisfied the selection as no disease has appeared in Kashmir from any of the eggs so supplied.

I feel certain that it would be a great mistake to make any change, more especially as M. Arbousset's sericicultural methods have enabled him to supply Kashmir with eggs at a much lower price than any of his confrères, French and especially Italian. I believe him to be a sericicultural authority thoroughly to be relied upon. No such satisfactory results have been obtained elsewhere and I should deprecate any change; but at the same time selections of other races might be tried for comparison, especially of Italian and crossed breeds. M. Arbousset informs me that if he were not also a reeler of cocoons as well as a breeder, he could not be as successful as he is, but thanks to his filature, he says, he can refine his cocoons without too great losses. This I fully believe. He informs me that last year (1902) he sent from his breeding grounds in Var 8,000 kilogrammes of cocoons for grainage to his filature in Alais.

I have had from the first a written guarantee from M. Arbousset for my own satisfaction that the eggs he supplies to Kashmir are free from disease.

How to obtain the greatest number of cocoons from the seed distributed.

First of all it is necessary to keep the seed properly sheltered from variations of temperature. Mr. Walton understands this perfectly. Then the worms should have plenty of space, particularly when they are young. The worms from 1 ounce of seed should occupy a surface of 4½ yards, that is, the worms from the hatching up to the first moult. They should be kept at a temperature of 23°C. (73°F.) to 25°C. (77°F.) and should have four meals in 24 hours and plenty of fresh air.

The right quantity of eggs to distribute in the work-people's houses.

The quantity should vary according to the size of the house and the number of the people in it. A family of a man and woman having one young child, could by working well, rear a box of seed of 30 grammes (one ounce); but they must arrange screens or drying poles so as to give to the worms, when they mount the branches to make their cocoons, at least 50 to 60 yards of space.

If the family consists of one or two older people of about 50 years and two children of, say 12 and 16 years, these could easily rear two boxes or 60 grammes of seed, under the express condition that they give a space of 100 yards to the worms when making their cocoons. The question of space is quite as important as the question of numbers.

Precautions in the Nurseries (magnaneries).

The first precaution to take is to make sure of a current of air, a good ventilation, and at the same time an even temperature of 22°C. (71°F.) to 23°C. (73°F.) from the hatching to the first moult and 20°C. (68°F.) during the remainder of the "education" (rearing) and five days after the worms have mounted the branches ("bruyère"), so that they can form their cocoons under the best conditions.

The litter should be removed two days after each moult, and twice, if possible, between the 4th moult and the mounting; the 2nd day after the 4th moult, and the 5th day afterwards.

To prevent loss of eggs.

Loss is preventible and should never exceed 5 per cent. M. Arbousset strongly advises a translation into the Kashmir language of his printed instructions and for each cottage to have it framed and hung up. These instructions I forwarded to Mr. Walton some time ago, and, no doubt, they have been circulated and acted upon, because the loss has not only been wonderfully less in the rearings of last year, but the average has been brought down almost, if not quite, to that of France and Italy where the rearing of silkworms is thoroughly understood and the loss of eggs during hatching does not, as stated above, exceed 5 per cent., which is very reasonable and small especially when compared with the loss of newer sericicultural countries where the rearing is not yet so well understood as I will show below:—

The produce of cocoons per ounce of eggs in France.

One ounce of eggs or 30 grammes should realize between 83 lbs. to 88 lbs. of fresh cocoons, equal to 49 lbs. to 52 lbs. of dry cocoons, or say, 1 kilo 250 grammes of cocoons from 1 gramme of eggs.

The cocoons from the fresh state to the dry lose 60 per cent. of their weight. 22 fresh cocoons equal 7.22 dried ones. In countries where the rearing is not yet so well understood, such as Hungary for example, they obtain only $2\frac{1}{8}$ lbs. of cocoons from 15 grains of eggs; in Persia and Central Asia where the rearing is very inefficient, they scarcely could obtain $\frac{1}{2}$ a kilogramme $\frac{1}{10}$ of pound of cocoons from each grain of eggs, whilst on the contrary, in the small silk-rearing districts of the Var and the Alps they obtain at least 2 kilos 500 grammes from each gramme of eggs hatched.

For further suggestions for rearers of silkworms see translation of M. Arbousset's printed instructions a copy of which I have brought.

Translation of M. Arbousset on losses of silkworms after the fourth moult, being an important letter in reply to some questions I lately put to him.

Your letter of June 3rd found me at Gonfaron (Var) during my busiest season in the hatching of silkworm eggs, and in the rearing of the worms, this will explain the reasons of my delay in replying.

I understand that the losses of which you complain arrive after the 4th mue (moult); they are generally due to a heat-wave, or what we call here "une touffe."

In order to escape this danger as much as possible, you must first put the eggs down to incubation as soon as the first buds of the mulberry trees begin to open, so that the harvest will be begun and, if possible, finished before the hot weather arrives.

Raise the worms to the perforated places in the doors and numerous windows, letting in a strong draught of air when the temperature becomes too hot.

Draughts of air are never injurious to silkworms. It is great mistake and a dangerous prejudice to believe to the contrary.

If the heat persists in spite of the draughts of air, water the roof if possible abundantly, just over the worms, but if that is not possible, hang large damp cloths, sheets for instance, over the place where the worms are, which by evaporation produces a rapid lowering of the temperature. But if you use damp sheets you must let the draughts of air continue, for in a closed room evaporation of the water by increasing the dampness of the air becomes a danger instead of a remedy.

These are, dear Sir, means for saving the silkworms from a heat wave if you care to add them to any letter you may be writing to Mr. Walton. I think you will be doing him a great service, and that you will be able to make sericiculture prosper in a district which is already indebted to you for so much progress.

Here in Var, where the climate is so mild, we have had a splendid harvest, which gives me confidence in the rearing stage, the moths are emerging full of life and activity in spite of the weather being rather fresh.

Yours, etc.,

LAURENT DE L'ARBOUSSET.

Ordering Graine.

Eggs must be ordered before the 15th of May, from M. L. De L'Arbousset, this he wishes me to impress on Mr. Walton.

I think the ordering of eggs should remain entirely in the hands, and at the discretion of Mr. Walton. I notice with much surprise in a report by the French Consul at Bombay, dated January, 1903, on sericiculture the following sentence, translated and sent to me by the India

Office: "French houses are advised to send cases of eggs through Latham & Co., of Bombay, to Mr. Peychaud, of Srinagar, who will transmit them to the local authorities to the Director of Sericiculture and to the Maharaja for his private estates." In my opinion this will be an unwise step. Any interference with the present method, which is the most direct possible, will be sure to lead to irregularities and confusion.

Non-exposure to light of Cocoons.

There is still a difference between the silk of the Cevennes and that of Kashmir in the paler colour of the latter, although since 1901 the colour has been much better.

One preventive is to avoid direct sun-light and day-light exposure both to the reeling machines and to the silk in every stage as much as possible. Their reels when filled by the cocoon reelers are transferred to low dark rooms. The yellow colour of the gum-silk is very evanescent to light exposure. A fine golden yellow raw-silk adds to its attractiveness to the buyer.

Also M. Arbousset thinks that in killing the cocoons by steaming the fault will be very much avoided. This is done by the Etouffoir which I will describe further on.

Cocoon-sorter.

This necessary instrument is already largely in

operation at Srinagar, and need not be here described. It was sent by M. Arbousset as well as the Etouffoir some time ago.

Cocoons of Reproduction or Etalons.

These can be supplied at 15 francs per ounce of 30 grammes.

If the project in buying them is to enable Kashmir to breed its own eggs M. Arbousset fears an unsuccessful attempt, as in supplying these it is not absolutely possible to have some of the eggs free enough from pebrine to obtain good reproductions, for to *succeed*, the country and the climate must be favourable to preserve the race, a risk in Kashmir not worth running. This was tried in Cyprus but failed each year, although the Cypriotes used the "graines selectionnès" of the Var. It has proved also the same in Syria where the shape of the cocoons grew smaller and degenerated each year to such an extent that every year they had to come back for cellular graine to France. The Royal Inspector wished to have a large quantity last year to regenerate the race, but could only be supplied to a quarter the extent required. It is better, therefore, to purchase the celluled eggs as Kashmir is now doing. The shape of the cocoons becomes modified even in the first year according to the temperature, the nature of the soil and climate, where the eggs of reproduction are introduced and it is difficult to attempt reproducing in any country where the people are not trained to reproduction. M. Arbousset tells me that he gives this information in order to prevent Kashmir making a mistake, but he is quite willing to supply them if it is wished. I am of opinion that it is not worth while to make the experiment, but to *let well alone* and go on as at present. If the natives can be trained gradually to all the *minutiæ* and scientific requirements then it will be well to try the experiment. But the lesson I learn from my visit and much correspondence and visits to the South of Europe, to the Government Sericicultural Stations at Padua and Montpellier is that as good seed cannot be grown in the Cevennes where the best silk is made, but the eggs have to be reared in the Department of Var, also that in India, *pari passu* the best quality of indigo is produced in the Behar district from imported plants, it is clear that one district may be suitable for silkworm rearing and reeling of cocoons, but that another is better for production of eggs. I do not advise any change. The advice given at the India Office Conference should be followed.

Production of silk from cocoons.

In France four kilos of cocoons produce one kilogramme of silk.

The production from one ounce of eggs realises between 83 lbs. and 88 lbs. of fresh cocoons, equal to about 49 lbs. to 52 lbs. of dry cocoons, or say 1 kilo 250 grammes of cocoons are produced from one gramme of eggs as stated in a former part of this paper. Fresh cocoons lose 60 per cent. of weight in drying.

On the respective merits of the French Etouffoir and the Dubiné Séchoir.

I had a good discussion with M. Arbousset on this very important question. He showed me that in France they prefer the cocoons to be Etouffées by steam on the Etouffoir. They believe that cocoons reel better than those passed through the Dubiné Séchoir if they are reeled say four or five months after and up to the end of the year, and they cost much less; but cocoons passed through the Séchoir reel better if reeled immediately and up to two or three months immediately following the Etouffage in the séchoir; but in any case Etouffage by steam or Séchoir shows considerable progress on Etouffage by the sun, or in other words for cocoons to be reeled immediately it is better to use the Séchoir, and for those reeled in four months and afterwards it is better to use the Etouffoir, provided the climate is sufficiently dry and hot to dry the cocoons at the end of August like that of France and Italy.

To steam cocoons in the Etouffoir or steaming machine they should be in it 20 minutes for the first time and then 15 minutes afterwards; thus passing cocoons through it four changes in each hour.

The Séchoir or drying machine has nothing to do with the Etouffoir. The Séchoir is a hot air machine whilst the Etouffoir is steam.

In damp climates, such as Tonkin and the south of China, where it is impossible to dry the cocoons naturally by putting them on shelves, it is necessary to use the Séchoir: but it is important to know that when the cocoons have passed through the Séchoir, if they are not reeled within the following three months the thread breaks oftener than when the cocoons are Etouffées by steam.

The Sale of Cocoons.

This is a subject I discussed at much length both at Lyons and Alais.

The fact of the filatures at Srinagar not having proved adequate to the requirements of reeling all the cocoons produced in each year and especially in 1902, owing chiefly to a much larger yield from the same quantity of eggs by improved *magnanerie* attention and system of egg-distribution which Mr. Walton has introduced, had caused me to think it might be advisable to enquire if any cocoons left over after the winter closing of the filatures could be exported to any country or countries where the filatures are open all the year, and knowing also that Marseilles was an important centre of cocoon commerce and distribution, I made the enquiries which result in the following information.

Whenever Kashmir thinks it advisable to export any cocoons to Europe they will be eagerly sought and bought at Marseilles, but upon the point of advisability I shall have more to say presently as the information I have obtained has greatly modified my opinions as to the policy of exporting cocoons.

The present time is an excellent one to sell cocoons at Marseilles. They are fetching a good price namely 10 francs 50 centimes per kilo (3/9 per lb.) for cocoons are scarce.

They need only to be simply sent to Messrs. Henry S. King & Co., Marseilles, who can easily dispose of them.

I saw Mr. King's brother in London and discussed the matter with him; he told me they would be glad to have the cocoons consigned to their Marseilles branch where they would be put on the market and sold to the best advantage; it is quite needless to employ any other merchants or brokers. To export cocoons the following precautions are necessary. They must be packed in a single strong piece of cloth as lightly as possible without crushing them. The bales must not exceed 80 lbs. to 90 lbs. in weight, for above that they are charged by 500 grammes and there would be a loss in carriage.

The following are names also of other Marseilles Buyers and Importers of Cocoons and Raw-silk.

W. Mooser, 6, Rue Nicolas, Marseilles.

Giraud Frères, 42, Rue Saine, Marseilles.

Stocker, Goldsmid & Co., 50, Rue Montgrand Marseilles.

S. D. J. Bensimor, Rue de l'Arsenal, 24, Marseilles.

Chabrières, Morel et Cie, Rue Paradis, 52, Marseilles.

L. Désgrand et Cie, Rue Montgrand, 14, Marseilles.

Alléon, Domerque, Rue Sylyatelle, 44, Marseilles.

Amade, Rue Montgrand, 40 Marseilles.

H. Estrangin, Place Paradis, 6, Marseilles.

As the distance from Kashmir to Marseilles is so great, it might happen that the constant rubbing together of the cocoons might damage them, and it would be better first to send a small trial lot, say about 500 to 600 kilogrammes uncrushed cocoons which always reel better than crushed ones: but some might be sent, say half-crushed also,. this would make a more economical transport, and 500 to 600 kilogrammes might be sent with the uncrushed ones, the results could then be compared.

M. Arbousset (if any should be sent) would ask a cocoon-reeling friend to put his reeling bassines at our disposal to make these trials.

M. Arbousset informs me that he has been selling cocoons to his colleagues in France and Italy, and says he is quite ready to give practical advice to Messrs. King & Co., who have fine warehouses in Marseilles for storage of large quantities. He says that cocoons are always sought after in Marseilles, and that the quantity annually passing through is very considerable and approaches three million kilogrammes of dry cocoons annually. If it is thought well, and should it be necessary to send trial lots of cocoons as now recommended, M. Arbousset will take 500 kilos. of each kind, crushed and uncrushed cocoons, and purchase them at the current rate, and then report results to Mr. Walton.

It is the French and Italian filateurs who buy the cocoons sold in Marseilles, but the greater part are reeled in the Cevennes. When the English Government took over Cyprus I suggested to Mr. Chamberlain that as I had found in examining cocoons for the Royal Commission of the Colonial and Indian Exhibition, those of the old race of Cyprus produced the finest quality of silk, it would be worth while to try to revive the industry and the old race then almost extinct. After several interviews with him he ordered a recommencement of the industry which has been very successful, and the cocoons are exported to the Marseilles market and sold there now at the value of £20,000 per annum.

I brought with me from Messrs. Guerin & Cie. of Lyons (Filateurs and Dealers) a sample of

Persian raw-silk made from cocoons they had bought from Persia, to show how well imported cocoons can be reeled. The silk is of perfect quality and very much better reeled than can be done in Persia.

From what I have learned during my visit to the south of France, I do not recommend the export of cocoons. In the first place there is little or no risk in keeping a surplus unreeled through the winter, proper care being exercised. They keep better in winter than in summer, I erroneously thought the reverse might be the case, and, secondly, there would be a considerable loss of profit in the difference between the sale of cocoons and the sale of raw-silk.

Cocoon-reeling.

For the third time during the last six years I visited the Cevennes (Department Gard, chief town Alais), for the purpose of a closer enquiry into all the *minutiæ* of cocoon-reeling and its subsidiary branches.

I was very kindly received by M. Laurent de l'Arbousset, of Alais, and also by M. Messac, of Messrs. Messac & Cie., who reel the finest silk in the Cevennes. This company reel their silk by the system "Tavelette Chambon" which consists of making the croissure by the friction of two separate compound threads from two bassines upon each other. The other system is known as the ordinary or Dubiné system of tavelette by which the croissure is made by the rubbing or friction of each bassine-thread upon itself. I brought working models of each system for trial in Kashmir and comparison with the system there which is that used in Italy. I shall be able to prove which of the three systems obviates most the defects in Kashmir silk complained of. The two French ones I will describe further on with drawings.

On the French precautions necessary in Cocoon-reeling and on the modus operandi generally.

The first requirement in raw-silk is evenness of thread, therefore, the reelers must always be trying to keep the same number of cocoons unwinding in the bassine, taking into consideration that the cocoon which begins to unwind has a stronger and thicker thread than it has towards its close.

Those cocoons which commence are called new cocoons, and those which are nearly unwound or reeled old cocoons. Four new cocoons-threads being reeled together of the best Cevennes cocoons should, if carefully enough reeled, give a thread of 10/12 deniers, but as it is necessary to use the old ones, half or three quarter unwound, the best way to obtain these 10/12 deniers size is to see that the reeler must reel five cocoons mixed, three new and two old. Cocoons should only be regarded as old when they have become sufficiently unwound as to be transparent enough to show the chrysalis inside.

The silkworm simultaneously always deposits two fibres, one from each side of its head, the single fibre is in France termed brin, the double fibre bave.

The average fibre of a single cocoon bave is 2½ deniers. A cocoon often gives three deniers at the beginning and hardly two at the end.

The "titre" of silk is the size of the thread. In cotton the fibre is expressed by numbers representing the number of metres contained in one kilometre, but in silk, the fibre is expressed in deniers which are old Italian weights equivalent to 52 milligrammes for one length of thread of 400 elles or 476 metres. In Lyons the fibre is calculated in this way with only a slight difference.

Defects in the Silk grown in Kashmir up to the present time.

(1) Tenacity or strength.

The compound thread, taking it altogether, is very regular, there are, however, places in it which are too fine. In twenty trials in one skein, there was an average strength of about ten grammes, the highest being 11·65 and the lowest (which was an exception) being 7·16. Such a breakage lessens the elasticity of the thread. An effort must be made to keep the thread more regular by keeping up the proper number of cocoons going to the tavelette. Saving, however, this one exception the strength is now good.

In reeling and testing the raw silk, the testing ought to stand seventy to eighty tavelles, that is, that a skein being reeled off for two hours should not break four times.

M. Arbousset found in two skeins of Kashmir silk I sent him to give an opinion upon, that, of the irregular threads in them, one of the skeins was reeled with five cocoons and was quite regular, whilst the other was reeled sometimes with four, sometimes with five, and sometimes even with six coccons, and was consequently of irregular sizes of thread.

Opinion expressed to me by Mr. G. Maurice, of Messrs. Veuve Guerin and Cie., Lyons. "Quality fairly good for Bengal silk, but too soft for a produce from French eggs. Winding irregular, some bales winding ninety to a hundred tavelles and others fifty tavelles. The bad winding does not come from gommèes, but from the great quantities of passage, called volle, which can easily be avoided at the filature. Cleanliness, fairly good, although many nibs could be avoided. Some samples sent to New York have been judged not perfect enough yet for use in that market. He believes that the silk can be much improved at the filature by subjecting the cocoons to warmer water and making the croissure as long as possible which will make the thread rounder and plus agregé."

I am quite in agreement with him on this *latter* point. I found both Messac and Arbousset gave a croissure of from thirty to fifty twists and with longer croissure; Mr. Maurice also said in a letter to me after I left Lyons. "I have shown the silk to several persons, and of ten who saw it eight of them said I cannot guess what silk it is, it reminds me of Bengal silk, but the feel and touch of it is more like French silk or Italian."

On this I emphatically remark that the idea of resemblance to Bengal silk is absurd, as it is produced by another species of worm and totally of a different nature of silk.

Copy of a letter from Messrs. Louis Desgrand and Company, very large Silk Merchants of Lyons, Marseilles, Crefeld, and Milan, to Messrs. Durant Bevan and Company, of London, and kindly lent to me by them.

Lyon, February 7/03.

Messrs. Durant Bevan and Co.

Dear Sirs,

We beg to acknowledge receipt of your favor of 5th instant.

Account-sale enclosed, please find account-sale of the remaining 7 Bales of Kashmir 164/170 amounting to £771 13s. 6d. to your credit. We beg to call attention to the question of winding.

Though progress has been made on that point, we still see that the silk is reeled "a bouts volants" instead of "a bouts noues" and the winding cannot be good. All good winding silks are now made "a bouts noues" and no doubt the reelers in Kashmir could easily be taught to reel in that way. The greatest and most useful progress would thus be realised, as winding with 93/100 tavelles the silk would easily find throwsters willing to work it, whilst with its actual poor winding very few will work it and then only at very high rates. The attention of the filatures should be called to this most important point, also to the evenness of the size. This silk can be much improved in regularity.

We have no doubt that with proper care and attention many defects could be obviated, and then the silk would be used with more favour.

Could you also obtain a greater proportion of fine sizes 9/11 deniers (average 10 deniers); our friends could use it in larger quantities but that may be a question of cocoons, as all cannot produce fine sizes.

See also if, by proper treatment, a greater nerve could not be obtained, this silk is rather soft, perhaps it is a question of water; but cocoons produced by French seed ought to have more nerve-elasticity in the thread.

Yours faithfully,

(Sd.) Louis Desgrand and Co.

Reeling by Power versus by Hand.

My inspection of the filatures in the Cevennes convinces me now that cocoon power-reeling has its advantages and must on the whole be preferred. It is generally used in Europe and gives more regularity and gentleness of action in drawing off the cocoon thread.

I do not see any other advantage of manual labour turning in Kashmir than that of the cheap labour there, but little power is necessary, a single 100 basin filature could easily be driven by a four horse-power engine, or even less, or by a water-wheel.

The power-driven reels are started and stopped separately at will by each fileuse when required by just moving a small lever by her side. No skill is required.

The cost of power-driven reels with the latest model bassines, including boiler and steam engine in France, is 500 francs per bassine. The Kashmir workpeople would be able to manage the new arrangement of bassines as well as the old, and I think it would be well to ask M. Arbousset to send, say, half a dozen, they seem very complete and are a great improvement on the older fashioned sort. Each fileuse can work four sets of cocoons on one reel and from one bassine with four tavelletes making four skeins simultaneously and working at a high speed.

Artificial lighting of reeling filatures.

M. Arbousset agrees with me that electrical light is not advisable, it is too costly, too irregular and does not give a sufficiently diffused light. He prefers and uses ordinary gas with the Auer Bec burner. These burners have each an incandescent mantle and are placed about three yards apart, seven feet high, between the two rows of bassines. M. Arbousset allowed me to bring one of the burners. I saw them very satisfactorily at work in the evenings. He told me that in winter they work three or four hours by artificial light and make as good silk as that reeled in daylight and that gas, with incandescent lamps, gives the finest light.

M. Arbousset's Criticism of the Electric Light.

It is not steady or regular enough, either the arc, or incandescent lamp. It varies too much in strength, one moment it is very bright and the next just as feeble. These variations tire the eyes of the work-people and prevent them from seeing the ends or even the threads of the cocoons. He thinks acetylene might be used with incandescent mantles where gas is not obtainable.

Ventilation of Filatures.

I think it would be extremely desirable to induce the reelers in Kashmir to work a few hours each day in the winter months, and for that purpose the filatures should be made warm enough and all fog and vapour got rid of.

In my dyehouses I accomplish this with a trunk of wood through which heated air is drawn from a stove over a steam boiler, and poured into the rooms, this dispels all vapour and is entirely satisfactory. I have brought a plan of it. M. Arbousset effects this by small stoves placed behind the reels, and by pipes of steam passing close over the reels, these warm the rooms and dry the silk as it is being reeled. Smaller steam pipes run parallel to them which convey steam to each bassine. If the Kashmir filatures are too big to be heated, the new ones should be on the French model which I can explain, and which can be done at a cost of 40,000 francs per 100 bassines; but for much less in Kashmir.

Fog and vapour can only be avoided in winter by elevating the temperature of the air inside the filature sufficiently, say to 60°F. or 70°F.

Temperature of Cocoon-Reeling.

To batteuse the cocoons boil the water and move the cocoons about for a few minutes to soften them and to find the ends when the cocoons are transferred to the bassines; the water in the bassines should be at the temperature of 70°C. to 75°C. (158°F. to 167°F).

In M. Arbousset's filature each fileuse batteuses her own cocoons, but in Messac & Cie's. filature a separate girl is batteuse for several fileuses, preparing the cocoons for them.

Wages of Fileuses.

Are 1 fr. 50 c. per day; work hours six a.m. to six p.m.

Explanation of Terms used in Sericiculture and in the Conditioning of Silk at the Lyons Conditioning House.

Purger.

Means that the cocoons must be cleaned after being treated with hot water by the batteuse; but not to soften them too much, on the contrary the cocoons must be treated, cleaned and reeled in the least possible time; thus they must be boiled and beaten ten minutes in a pan full of water heated to 90°C. (194°F.) to boiling "*cuite.*"

Then (the cocoons so treated), bring the water to a temperature of 60°C. to 70°C (140°F. to 158°F.) by the addition of cold water, rapidly clean (Purger) the cocoons as they find the threads and put them to unwind by keeping up the temperature of the water in the bassine during the reeling. In this way any careful reeler can produce raw-silks of the same ordinary value of those of Europe, China or Japan.

Titre.

The titre indicates the size or number of the thread.

Bouchons.

"Peu de bouchons" means that the thread of the silk is sufficiently regular, clean or smooth, and has none of the little loops or nibs on the baves or thread called *bouchons*, or the still smaller ones termed duvet.

Tavelles.

The ascertaining by the Lyons Conditioning-House, say 85 to 90 tavelles, shows that a work-woman can reel alone in two hours 85 to 90 skeins or flottes of silk. This is a good average, for the very best reeled silks have not more than an average of 90 to 100 tavelles. If one skein or flotte has broken four times in two hours' testing for one fileuse 40 to 45 tavelles, of 27 turns and making 1½ per cent. of waste in winding 50 metres, this is considered bad, and too much below the average, as stated above.

Vrille.

This is a defect shown when the silk thread is twisted like a corkscrew.

Aggregation.

This is the putting together of several threads of the cocoons so as to form the silk threads, the best aggregation is obtained when the silk is properly reeled and the croissure is properly made whether it be by the system Chambon or the ordinary methods.

Venures.

Are formed when bouts are reeled, which have commenced with the first and enveloping layer of the cocoon being less coloured than the lower layers, thus giving veins or venures of paler silk in such parts.

Vole.

Vole is the silk thread that is too fine and has not sufficient resistance. It is produced when the fileuse does not keep all her cocoons threads going, for instance, when she is only reeling two cocoons when she ought to be reeling four.

Gommures.

Is the glueing of the silk fibres to each other. It is principally produced in damp weather, at the point of contact with the arms of the reels.

Elasticity.

The capability of the thread being properly elastic.

Tenacity.

Or strength; when the cocoon threads are not sufficiently aggregated together, the silk is considered a little tender, because the pull is on a part and not on the whole thread. In France the silk should not break at less than with a weight of 18 to 20 per cent. for 9/11 deniers. Three trials of one of the Kashmir skeins last year broke at 25 grammes, whilst others were 45 grammes. The Cevennes well-reeled silk breaks at 30 to 40 grammes.

The Yellow Colour of the Silk.

"Un peu veiné" means that on the golden yellow silk there are paler threads that form veins or venures with less colour than the others. This is objectionable, although the fault is not very serious. It can be avoided at the time of reeling by mixing the cocoons half-reeled with those that are only just beginning to be reeled. See venures.

Usages de Lyon.

Means that when silk is sold subject to the "usages de Lyon" it is subject to several expenses and conditions which amount to a discount of five per cent., for instance, the price of 40 fr. per kilogramme net is the same as 42 for "usages de Lyon."

To explain more fully, the official rates of Lyons silk are divided up as follows, say for one kilogramme of silk of a nominal value of 40 francs per kilogramme:—

Commission on sale, 1 per cent. ...	0·40
Del credere (responsibility of agent) 1 per cent.	0·40

Brokerage ½ per cent.	0·20
Discount on 90/100 days 6 per cent.	0·65
Deduction ½ per cent.	0·25
Half the expense of conditioning ...	0·75
Carriage from Alais or Marseilles to Lyon	0·10
	2·75

This explains the difference of two francs a kilo in the price of 40 francs between the net price and the price "usages de Lyon." When the seller sells at the net price, he pays the price agreed upon, but when he sells at the old "usages de Lyon" the deduction is made of all the above articles. The discount increases with the price, on 40 fr. it is 2 fr., on 50 fr. it is 2·50, on 60 fr. it is 3 fr., and on 100 fr. it is 5 fr.

Va and Vieune.

Is a wire covered with glass tubes which constantly moving from right to left in front of the reel distributes the silk threads, say two to three inches in breadth on the reel, and evenly, preventing the threads falling all in one place on the reel.

Guindre.

Means the reel on which the silk is wound from the bassine. In France it is placed behind the reeler. The reels are covered with cloth which receives the silk and prevents it coming in contact with the wood of the reel.

Blaze.

Means short lengths of refractory cotton-like short fibres from the outside of the cocoon which when found in déchets or waste are objectionable as they will not take the dye.

"*Vice rédhibitoire.*"

This is a Lyons term, and is used to describe the damage done to raw-silk by mites (dermestes) which get into the bales sometimes when they have been long in stock. They probably come from wool which is often infected by them. They gnaw the silk and make it unwindable. When this occurs the buyers can claim from the merchant or broker for any damage done by them. But before the buyer can claim upon or refuse the silk so damaged, there must be numerous skeins or flottes gnawed by these mites. To avoid this damage, do not allow the silk to remain too long in the warehouse and avoid in packing the use of any woollen materials, or any thing not clean; only use absolutely new cloth. Avoid also contact with bales of wool which always contain great numbers of these mites. Any remedy if applied is worse than the disease, and prevention is the only cure.

Quality of Silk to be aimed at in Kashmir.

It would not be wise to aim at the highest quality obtainable by several of the best Cevennes filatures. They chiefly produce 9/11 deniers raw silk for very special purposes and get proportionately special prices, but this is only for a very limited consumption. These silks are called extra and Kashmir would try in vain to get such extra prices. It would indeed be possible to arrive at this extra quality and extra price, but there would be more loss in damaged cocoons than would be gained in the price. If Kashmir will content itself with obtaining silk of the second order of France and Italy it will meet with more demand and make more profit; of this I am convinced. It is very close upon that at the present time. What is required to accomplish very desirable results is to make any necessary modifications here suggested, to encourage greater skill in the reelers and to have thorough supervision of the reelers, especially the younger ones, who are apt to be careless.

I noticed in the filature of Messac and Cie which I visited in Alais, that this firm was making this extra quality, whilst that of M. Arbousset was occupied with that of the second order and most in demand. I have brought small samples of Messrs. Messac's to show. They were kind enough to give them to me.

M. Arbousset's expression to me of his opinion of Kashmir reeling and raw-silk of last year's production (1902).

"I find an immense progress in the raw-silks of last year, they are clearer and more brilliant and present themselves in a much better way than the two bales Mr. Walton sent me to sell last year, which notwithstanding realised such a good price.

The cocoons should be reeled up to the end of their exhausting-point, not trying to reel double cocoons, these can be sold for 3 fr. 50 c. per kilo. just as they are in the dry state.

The waste-silk requires a little more selection and arrangement."

In November, 1902, M. Arbousset wrote me as follows: "I think highly of the Kashmir silk of the last few months."

White Cocoons and White Silk.

Enquire of Mr. Walton how the 433 ounces of the race of white cocoons which Arbousset sent him in 1901 turned out as to quantity, disease, &c., as compared with the yellow race, and what prospects he thinks there are for producing more white silk, this kind being much in demand in Lyons for silk muslin in white and pale colours just now. Discuss this matter with Mr. Colvin and Mr. Walton carefully as there is no doubt white raw-silk is now a desideratum.

Indian, French and English Weights.

One Kashmir maund is ... 82 lbs.

One French ounce weighs ... 482 grains.

To convert grains to grammes multiply by 0·0648.

One ounce of Silkworm seed weighs 30 grammes.

Purchases of Eggs from M. Arboussel, September 10, 1901.

Eggs purchassed in 1900 for breeding in 1901. 25,542 boxes of 1 oz. each, and 42 boxes in excess of the order.

The price is four francs per ounce.

Purchases in 1901 for 1902 crop.

25,542 boxes of eggs of 1 oz. each of which 25,000 cost £4,000, or 4 frs. per oz., he also sent 42 boxes in excess of the order.

In this consignment there were 433 oz. of a good race of white cocoons of Haut Cevennes.

Crop of Raw Silk, 1901—1902.

Mr. Walton has sent to Messrs. Durant, Bevan & Co., 553 maunds of silk; 400 more are to follow, he also sent 100 maunds to the local markets, making a total of 1053 maunds. This quantity, he says, he has obtained from 20 maunds, roughly, 20,489 oz. of seed.

Species of Silk Moths.

The silk moth whose larvæ produce the European, Chinese and Japanese silk of commerce is Bombyx mori.

The Barapaloo	of Bengal	is	Bombyx	textor.
The Cholopaloo	„	„	„	fortunatus.
The Desi	„	„	„	„
The Madrassi	„	„	„	cræsi.

The incubation of Bombyx textor is about ten months.

The incubation in Bombyx fortunatus is eight to ten days. In Mysore the Bombyx sinensis is cultivated. The Bengal worms feed on the Morus indica. It is shrubbed, not cultivated into trees.

The November bund of Bengal silk is the silk of the Bombyx fortunatus or Desi worm, that of the July bund is the Madrassi or Bombyx crœsi. The silk of this season is not quite so good as that of the November bund.

Tenacity and Elasticity of Kashmir Silk compared satisfactorily with Japan Silk in 1902.

The average of 20 trials of Kashmir sewing silk, which I made in Leek, gave a tenacity or breaking strength of 1 lb. 15 oz. $13\frac{3}{8}$ drams, that of Japan of the same size, viz. three-fold 20's was 1 lb. 15 oz. $1\frac{5}{8}$ drams; this is my own testing.

Mr. Kershaw on Kashmir Silk.

Mr. Kershaw is one of the largest silk mauufacturers in Macclesfield.

He informs me that he is much pleased with the silk and would like two bales specially reeled for him of the very best quality when I get out to Kashmir, one of $\frac{10}{12}$ deniers, and one of $\frac{13}{15}$ deniers and that he will probably require a minimum of 50 bales per annum. He also very much requires white silk to replace that of Japan of which he is using a very large quantity, and asked me to enquire if he might depend on a supply.

T. WARDLE.

CHAPTER XV.

MY SECOND SERICICULTURAL REPORT, WRITTEN IN SRINAGAR, APRIL 27TH, 1903, BEING SERICICULTURAL CONSIDERATIONS AND DISCUSSIONS ON THE FRENCH METHODS OF PROCEDURE AND SUGGESTIONS CONTAINED IN REPORT No. 1, PUBLISHED IN JAMMU IN MARCH, 1903.

THIS second Report is mainly the outcome of careful discussion and comparison with Mr. Walton, the Director of Sericiculture of the Srinagar Filatures, on the points raised in Report No. 1, consequent on my visit to the sericicultural districts of the South of France, where the best raw-silk is produced, prefaced with notes on the proper cultivation of the mulberry tree.

In perusing my present Report it will be necessary to consider in connection with it what I have described in Report No. 1 under each separate heading, here repeated for easy reference.

I have written a chapter on mulberry tree cultivation both as to the state in which I found it in Kashmir, and also as I think it might and ought to be. I venture to commend to the State the high importance of this subject, which, in my opinion, is so great and so pressing that I feel I ought to give it precedence in the remarks I have now to submit.

On Mulberry-Tree Cultivation.

I have been as much astonished as delighted to see the vast quantities of mulberry trees in the Vale of Kashmir. I have noticed that the prevailing species is Morus Nigra or the black mulberry. I would recommend also the cultivation of Morus Alba or the white mulberry, which is found to give the best results in sericiculture in both Italy and France. Mr. Walton informs me that the white mulberry is found in Kashmir; but probably not to a greater extent than about five per cent. of the black mulberry growth. He tells me the best results are obtained from a fruitless variety or species. The rearers use its leaves where they can obtain them for the last stage of silkworm feeding.

I noticed in my journeys through the villages, small plots or patches of mulberry plants, and on asking why they were so planted I was told that the Maharaja of Kashmir had issued orders for each village to provide a small mulberry-plant nursery. Nearly all of those I saw lacked supervision, and I cannot help thinking it would be well if an expert Mulberry Forest Officer were appointed to attend to this most important branch of sericicultural industry.

In examining the mulberry trees over a considerable area of the Kashmir valley, it is painfully evident that they are having little or no correct botanical attention, being of very irregular and in too many cases of worn-out growth, and requiring capable botanical oversight, either in the growth of the tree or in properly pollarding and pruning, as is the case in Italy and France. Instead of which, the trees are hacked in a most ignorant way, and are being so treated at this time of year when the sap is rising. This, I suggest, should be done under proficient European oversight notably by one of the staff of the Indian Forest Department, with a qualified assistant well acquainted with the methods and procedure of France and Italy. The contrast between the careful culture in these countries and the neglected state in the Kashmir valley is most striking. I have seen in the valley, hundreds of trees only fit for firewood. It cannot be expected, nor is it possible for silkworms to be properly fed and nourished by the leaves of such trees as these. In addition to this suggestion, I urge that no time should be lost in extending the area of mulberry cultivation throughout this and the neighbouring valleys, so as to provide leaves for villagers, who have now, in some instances, to fetch them from too great a distance, as well as to those who have not yet shared in the industry of silkworm rearing, and further, what is of great importance,

PLATE 10.

to give the silkworms a supply of young and succulent leaves instead of those from the old and in too many cases worn-out trees, now existing.

Since I wrote these lines I have been over a considerable area of mulberry tree growth, principally from Avantipura to Arapal, or from Kulbend to above Arapal and even as far up the valley as Narastan.

There are three thousand mulberry trees, chiefly very old and almost denuded of branches. The mulberry chiefly predominates over the willow, the chenar and the walnut (which is abundant and unmolested), but the mulberry is in such a state of devastation as to be incredible except upon actually seeing the remnants of what were once mulberry trees of splendid growth; but now almost denuded of leaf-bearing boughs.

The most valuable parts of the trees have been cut away with axes in order that the leaves could be gathered on the ground, in many cases for firewood, some of it quite recently, and also needless and cruel lopping.

I have spent several days in this locality with Colonel Ward who is equally astonished at the destruction everywhere too evident. He fully supports my statements.

In France and Italy the leaves are gathered by ascending the trees, or by ladders without cutting down the boughs, and no leaf-bearing part is damaged.

If this deplorable state of things is permitted to go on for the next four or five years there will be no mulberry leaves to gather. The use of the axe should positively be prohibited. I have never witnessed a case so illustrative of the fable—killing the goose for the sake of its golden eggs. This state of mulberry forestry is most deplorable and should not be continued, for there can be no difficulty in gathering the leaves, although many of the trees are high the Kashmiris are an agile race and it is wonderful to see them climb a tree; besides other methods than that of the axe could easily be adopted. The present system is wilful destruction not only of the mulberry trees but of the silk industry and calls for immediate reform. In all my experience of forestry in Germany, France and England I have never seen such a deplorable state of devastation. Heavy penalties should be enforced wherever people are found lopping off boughs and all instructions for pollarding, coppicing and pruning should be carried out by disinterested and experienced persons, and only at the proper season.

With regard to pollarding the existing mulberry trees, it must not be understood in the sense of pollarding in France and Italy. There the case is widely different. The pollard trunks are grown for the purpose of producing pollarded branches from their tops, as with the pollard willows of Kashmir. The branches are consequently and generally much smaller than the boughs of the large Kashmir mulberry trees, which for the most part are fine trees of indigenous existence, self-raised and grown into very large trees in many parts of the vale, with powerful boughs and branches, where they have been left intact.

These should be retained as far as possible to allow the younger leaf-bearing shoots to sprout from, and not lopped off in the present wholesale fashion leaving the trees mere skeletons.

The demand for silk is yearly increasing with the increase of population, particularly in the United States of America, now the largest silk textile manufacturing country in the world. It is almost certain that as soon as Kashmir raw-silk is raised by better reeling to that of the best qualities of Japan silk now chiefly used there, it will be greatly required and the enquiry for it enormously increased.

What I have suggested with regard to the planting of more mulberry trees and on improved cultivation generally, ought not to be delayed as several years must elapse before the younger leaves can be gathered, and it is quite possible, and very likely, that the demand for Kashmir silk both in Europe and in America may exceed the supply of mulberry leaves.

No. 1.

On Cellular Seed.

The eggs or seed supplied to Kashmir by Messrs.

Arbousset & Son, are of the kind of race described under this head, in my report No. 1, but sent in perforated paper boxes for greater protection. They are labelled "Graines de Vers A Soie Cellulaire Selectionnèes des Montagnes de Mauriers. Preparèes Selon le Method Pasteur."

Mr. Walton informs me that he finds very great freedom from disease and that he is quite satisfied with the eggs purchased and which are always guaranteed free from pebrine. The precautions described are necessary to be observed in Kashmir, and great care has to be exercised to prevent the natives bringing in cocoons to the filatures before the silkworms have turned into chrysalides. This they attempt because their labour in rearing the worms from the hatching of the cocoons is paid for by weight, and worms are heavier than chrysalides. The remarks on this important subject in Report No. 1 have been well considered by Mr. Walton. They have his full concurrence, and so far as the instructions therein contained as to silkworm rearing in Kashmir, they are acted upon as thoroughly as is possible.

No. 2.

How to obtain the greatest number of Cocoons from the Seed distributed.

The following subjects under headings No. 3, 4, 5, and 6, being so closely related to that of No. 2, I have classed them with No. 2.

No. 3.

The right quantity of Eggs to distribute to the most people, i.e. the village-rearers' houses.

No. 4.

Precautions in the Nurseries (Magnaneries).

No. 5.

To prevent loss of Eggs.

No. 6.

The Produce of Cocoons per ounce of Eggs in France.

These are collateral subjects, the remarks upon which in Report No. 1 have been well noted by Mr. Walton. I have discussed them seriatim with him and there is no necessity to enlarge further upon them. I find them well understood and practised here.

Mr. Walton tells me he obtains from careful rearers quite as many cocoons from the same number of eggs as are obtained in France and Italy, and in some cases more. Last year his average was 39 kilogrammes of cocoons per ounce of eggs, against 36 kilogrammes last year in France, but the yield of cocoons must always be variable, owing to climatic conditions and other reasons.

With regard to trying different races of eggs, Mr. Walton promises me that he will act on my suggestion of trying selections of Italian races.

The high price of Italian eggs is against any

violent change, and so long as the silk produced in the Cevennes enjoys such a high reputation no anxiety for change need be entertained. The confidence in M. Arbousset's great sericicultural knowledge and trustworthiness is well bestowed. He and his son have served Kashmir well.

No. 7.

M. Arbousset's letter on losses of Silkworms after the fourth moult.

Mr. Walton has carefully considered this letter and is in agreement with it with one exception. He says that the climate of Kashmir does not admit of exposing the worms to the strong currents of air recommended in this letter. He has tried the idea but lost many silkworms, because when a shower of rain follows a very hot day the temperature is lowered to such an extent that the worm dies, no doubt from the close proximity of the snow-clad mountains. If the temperature falls from 70° to 50° the worms die. Regulation and evenness of temperature must be maintained in the Magnaneries, and for that purpose Mr. Walton has provided all the rearers with thermometers.

No. 8.

Ordering of "Graine" or Eggs.

Nothing more need be written on this head than to emphasise what I wrote on page 179 in

Report No. 1 and to commend it to the serious attention of the State.

No. 9.

Non exposure to Light of Cocoons.

It is admitted that the practice in France should be followed, which is to avoid direct sunlight and daylight exposure of both cocoons and the reeled silk, as much as possible. The golden yellow colour of the silk is very evanescent.

Mr. Walton has discontinued the sun-drying of cocoons, except in the Lolab and Üttermuchipore districts, 40 to 70 miles from Srinagar. These districts are so far away that if the cocoons were to be carried so far they would not only get crushed but the moths would emerge from them. The only other way to provide against this contingency would be to provide these places with séchoirs, not at present advisable. In other distant districts like Khanabal, Achabal, Vernag and other far-off valleys, the case is different, because the cocoons are brought by boats down the Jhelum to Srinagar in less than 24 hours.

I think the present filatures are quite light enough to permit of reelers to do their work properly, but any excess over this should be avoided.

No. 10.

The Cocoon Sorter.

This useful machine answers its purpose admirably

No. 11.

Cocoons of Reproduction, or Etalons.

The meaning of this is that eggs might be imported from which in the following year silk-worms could be bred from eggs hatched here and so prevent the necessity of importing eggs annually. This method was recommended by one of the speakers at the India Office conference at which I presided. I strongly demurred to this opinion and pleaded that we should let well alone. It is an erroneous opinion and has proved so by experiments during two years here. The principal reason and secret of the success of sericiculture in Kashmir, has been and still is, in annually importing fresh and healthy eggs from the best European egg-breeding districts. As may be seen in the India Office Conference Report of the 26th September, 1901, from the first I stated that I had recommended Mr. Walton to try eggs of reproduction, but only experimentally. The first year 3,200 ounces of eggs of reproduction *were* bred in Srinagar, the result in cocoons being almost nil. The eggs hatched out very well, and up to the third moult there were no losses, but after the third moult the worms died, some from pebrine and more from flacherie. The second year 6,400 ounces of eggs of reproduction were reared locally from seed at first imported from Italy and France with a result of only 200 maunds of very inferior cocoons. If they could have been reared free from disease there would have been a crop of 6,400 maunds of cocoons. This branch of sericiculture is much too scientific in several ways to be attempted for a long time to come in Kashmir. The remarks of M. Arbousset on this subject on page 181 of my first report are quite conclusive and need to be well kept in view. It is not only in Kashmir that the experimental efforts of buying cocoons of reproduction have proved a failure. It was the same in the Lister Dehra Dhun grant on a large scale and in other efforts to promote sericiculture elsewhere. Lord Masham in congratulating me on the Kashmir success, wrote to me that he had lost £50,000 and given the undertaking up. I am distinctly of opinion that the present system of importing eggs of the best races should be continued. It is really the secret of Kashmirian success.

No. 12.

Production of Silk from Cocoons.

Mr. Walton tells me he has succeeded in raising his production to 1 kilogramme of raw-silk from 4½ kilogrammes of cocoons. This is very good; but he says the rearers do not always carefully feed the worms and are often not sufficiently careful in other respects.

The French Reelers only buy selected cocoons and in this way get a higher percentage of silk than is possible in Kashmir, where all kinds of cocoons have to be taken into the filatures including double and inferior ones.

From 40 kilogrammes of dry inferior cocoons it is not possible to obtain more than about 6 kilogrammes of silk.

No. 13.

On the respective merits of the Etouffoir and the Séchoir.

On this important question I find difference of opinion. See Arbousset's statement in report No. 1 page 183.

Mr. Walton says he was aware of M. Arbousset being in favour of the Etouffior at the time he had only one Etouffoir and one Séchoir.

He, Mr. Walton, is distinctly of opinion that the Séchoir has given him the best results, and for this reason, that in a humid climate like this, the cocoons when passed through the Etouffoir become mildewed by retention of the condensed steam. The atmospheric moisture of the valley prevents their drying sufficiently after the Etouffage. Also that in the filatures there is a want of space for drying such an immense quantity of cocoons. This inherent dampness causes the thread to become tender, and besides that, cocoons having to be turned several times a day to expel moisture cause disorganisation of the fibre and more or less crushing the cocoons, making difficulty in reeling.

Mr. Walton says it takes two or three months even when so often turned over after Etouffage, before they are dry enough to reel and even then mildew sets in.

It is well to observe that, notwithstanding the merits of Etouffage and its prevalent use in France that even M. Arbousset endorses Mr. Walton's opinion, where he refers to the damp climates of Tonkin and the south of China, see Report No. 1 page 184.

The dampness of climate in Kashmir, caused by the rainy season, the great floods, the great irrigation in the valley and consequent evaporation, happens chiefly when the cocoons have to be dried when the humidity of the atmosphere is from 90° to 100°, saturation being 100°.

The humidity is daily registered both in the filatures and in the excellent Hibernation houses at Gagribal, some distance away, by the ingenious instrument called Lambrecht's Polymeter.

It goes without saying, that the less cocoons are pulled about the easier they can be reeled and with a superiority of quality and freedom from duvet and bouchon, the latter faults being the principal defects now existing, so far as my observations go up to the present time of my visit.

Although I have been and still am of M. Arbousset's opinion, formed from the experience gained in several visits to the South of France, I feel bound to admit that the foregoing reasons prove that the expenditure of adding more séchoirs, necessary for killing and partially drying the cocoons, is a proper and defensible one. Mr. Walton finds that it is better to only three-parts dry the cocoons in the Séchoir, and he confirms M. Arbousset's statement, that if the cocoons are dried too much, the thread breaks more easily.

If the atmospheric conditions here were those of the South of France, where the humidity is probably not more than 40 to 60 degrees and where there is scarcely any irrigation, I have no doubt it would be preferable to treat the cocoons by Etouffage and that mildew would not set in. When cocoons are taken out of the Etouffoir, they are of course quite wet from the steaming and it is clear that if they cannot be quickly dried mildew will naturally supervene.

On my return through France I will discuss these points with M. Arbousset and report further. I may be now permitted to observe however, that if adequate means were provided for drying the cocoons at once after steaming the French system might prevail and prove the best. My only anxiety is to raise the quality of the raw-silk up to that of the Cevennes.

No. 14.

The Sale of Cocoons.

I commend my notes on this subject see page 184 of Report No. 1.

I have spoken about this subject at length with

Mr. Walton. He quite appreciates the importance of the discussion I had with M. Arbousset upon it and argues that although it would be unwise to get into the habit of sending cocoons to Marseilles for sale, instead of reeling them at home, it ought only to be done when the price of cocoons is high, say not less than 11 to 12 francs per kilogramme, as at present; cocoons being scarce and much in demand in France and Italy to keep their reelers employed during the winter; but he quite agrees that Srinagar cocoons should only be exported when the surplus stock left over from the previous crop cannot be conveniently reeled before the new crop is ready for reeling, as has been the case with a large portion of last year's crop, in fact about two-thirds.

The fact however of four more filatures being in a sufficiently advanced state of erection to be ready in June, to receive the new crop, it is hoped that surplus stocks will be reeled in future by the end of the following June, at which time the new crop is ready for reeling. I am very glad to learn that silk-worm rearing and cocoon producing are becoming more popular and wide-spread in the neighbouring villages and valleys, and also are very remunerative to the peasants; there will consequently be a liability to surplus stocks unless the increase of filatures keeps pace with the ever increasing production of cocoons.

The following table shows the yearly increase

since my first purchase of eggs for the Government of India in June, 1897:—

Year.	Quantity of Eggs in ounces imported.	* Ditto Local.	Value of Eggs and Carriage.	Raw Silk produced.	Waste Silk produced.	Local Silk & Waste Inferior Cocoons.	Total Value of Out-turn.	Number of Rearers who took eggs.	Average Number of Rearers engaged.	Quantity of Fresh or Green Cocoons.
			£	lbs.	lbs.	lbs.	£			lbs.
1898	1,920	...	766	5,412	1,863	351	4,858	326	1,304	1,07,100
...	...	6,400	...	Nil	Nil	Nil	...	1,106	4,424	Nil
1899	6,400	...	1,632	22,509	7,500	7,239	...	803	3,212	3,26,672
...	...	12,800	...	560	320	500	14,684	2,255	9,020	16,000
1900	19,060	Nil	3,000	44,181	23,063	7,739	27,419	4,290	17,160	8,13,792
1901	25,606	do.	4,333	65,931	32,870	12,346	49,118	5,887	23,548	10,65,204
				Estimated at						
1902	25,527	do.	4,333	†1,20,750	†54,500	†12,000	†96,666	†8,158	32,632	†18,82,692
1903	25,521	do.	4,333	‡ ...	‡ ...	‡ ...	‡ ...	11,060	44,245	...
1904	35,500	do.	5,500	...	...	...	...	...	...	...

* The Eggs were of local reproduction, both moths and worms died of Flacherie and Pebrine.
† The exact figures to be obtained later on.
‡ Not yet realized but anticipated to be very satisfactory.

The great advantage Kashmir possesses in the present price of labour is one that puts the south of Europe sericiculturally out of competition, especially when the silk in reeling shall be brought up to a higher standard and the reels can be run as fast as they are in France, which is quicker than that of the Srinagar filatures and doing better work, the difference in cost being about three times cheaper than in France; as I shall prove further on.

With regard to the sale of cocoons I may say further, that the question of the employment to the fullest extent of the spare labour of Srinagar, of which there is yet, as I am informed, no paucity, should have prior consideration, because the people are exceedingly poor, some of them being on the verge of starvation. It will be conceded, I think, that this is even of much greater importance to the State than gaining profit by the sale of cocoons.

Should it be ever necessary to export cocoons, the information I have given in Report No. 1, pages 184 to 188, will be a safe guide.

No. 15.

Cocoon-reeling.

I have seen trials made with the two tavelettes I brought from France, the ordinary or Dubiné tavelette, which I find is the one I introduced from Italy in 1897 and which Mr. Walton continues to use, and the Chambon in which the croissure is made by the crossing of two separate threads and not by one, as is requisite for extremely superfine qualities of raw-silk, which do not represent the staple output of French and Italian raws.

I found a difference in the distance between the bassine eyelet and the top tavelette pully between which the croissure occurs; the distance in the Kashmir tavelettes, copied from the Italian method, being a few inches, at least six, shorter than in the French method.

In my opinion the French method is better than the Italian and is now being tried here.

Mr. Walton gives more croissure than the French, the French being 30 to 50 crossings and Mr. Walton's 60 to 100.

The latter is probably necessary, because the French cocoons, being selected, may not need so much croissure as those of Kashmir. I think it would be better to sort the cocoons.

No. 16.

On the French precautions necessary in Cocoon-reeling and the modus operandi generally.

Nothing need be added on this head. I find everything I have stated in my Report No. 1, pages 189-190, is well understood and practised in Srinagar.

No. 17.

Defects in the Silk grown up to the present time.

Tenacity or Strength.

It will be necessary to refer to the faults found

as stated in Report No. 1, pages 190-3, to prevent repetition here. Efforts are being successfully made to evolve the thread so that it is more regular by keeping up the proper number of cocoon threads going to the tavelette. It is only by unintermitting care by the reeler and the closest supervision over him that a permanent improvement can be effected. The defect arises solely from careless preparing and especially reeling, and also from inexperienced hands allowing one or more threads to be dropped out, thus making in places too fine a thread, which of course reduces the tenacity. I feel sure this defect is being gradually remedied. Even during my stay the average of the testings has been raised.

Mr. Walton always conducts this important operation of testing personally and does not permit any one of his staff to do it. I have watched a number of his testings and find the average has been raised from 50-70 tavelles to 80-90 and some reach 90-100.

A higher average than even 80 or 90, which is fairly satisfactory, can be obtained by more thorough and adequate supervision, in which case the value of the reeled silk would be raised by 1s. per lb.

To obtain this higher excellence of the best French and Italian reeling, I am certain that there should be one European superintendent to each filature and still better with two, and with a proportionate number of thorough and well trained native assistants. Let me inform the State, that in France I found one elderly man over-looker of great experience in every filature of 100 bassines, assisted by two elderly females, who themselves had been reelers all their previous working lives, and who were highly qualified to superintend the 100 fileuses, 50 on each side the filature. Here, on the contrary, I find two of the filatures with 424 bassines controlled by only one European but with fourteen native assistant overlookers. Whilst I strongly recommend a minimum employment of Europeans and a maximum employment of native overlooker, I feel sure that it is not paying policy to stint either European or native supervision.

Let the boys be encouraged as much as possible to take pride in excelling in their work.

These are my own impressions which I offer for whatever they may be thought to be worth, both to Mr. Walton and the State, but I recommend their consideration to the serious attention of the State; suggestions which if adopted more or less would materially increase the value of the silk, and would do away with the brokers' and manufacturers' complaints, which I have enumerated on pages 190-3 of my Report No. 1.

These complaints I know to be well founded, although often exaggerated by the manufacturers in order to get the silk at a lower price.

I proved the truth of some of these exaggerations by recently having a quantity of raw-silk of Srinagar reeling, thrown, dyed, and manufactured into brocades. These I have brought over to show the Maharaja, Raja Sir Amar Singh, and the Durbar. They are quite equal in quality to ordinary French brocades, and were much admired at the Durbar held at the Palace at Jammu to receive me. Two lengths, for dress purposes, are being offered to the Queen and the Princess of Wales by the Maharaja. Notwithstanding this, the raw-silk is not quite equal to what it may, and, I doubt not, will be, if Mr. Walton is well backed up in everything he may think necessary.

Specimens of these brocades may be seen in the Sir Pratap Singh Museum here where the Maharaja has recently sent them.

In a letter from Messrs. Louis Degrand & Co., in my Report No. 1, pages 192-3, it is stated that the reeling should be "à bouts nouées" and not "à bouts volantes." The former means, that when a cocoon-thread breaks the reeler should tie it together by a knot. "A bouts volantes" means that instead of tying the thread, the reeler to save trouble throws the thread upon the reel instead of piecing it by a knot to the other broken end.

I find that it is forbidden to allow "bouts volantes" and it is only by the closest surveillance that this carelessness, which is here termed "budzati" can be avoided. It takes a little trouble to find the proper end, which the Kashmiri reeler does not like. Only incessant watching will prevent it.

No. 18.

Reeling by Steam or Water-power versus Hand-power.

This is an important question, in which Mr. Walton and I are not in perfect accord.

It is well to give his opinion and reasons. He says, that many years ago Messrs. Louis Payen & Co., who have extensive filatures in Bengal, erected a factory of 200 bassines on the French principle of steam-power turning. The quality of the silk turned out proved to be inferior, so much so, that they had to re-reel the silk before sending it to Europe, also they found that by giving up hand-turning by boys, they could not increase the number of workmen, as can be and is at present done by employing juvenile hand labour, which is daily gaining experience by seeing the reeling operations, in helping the reelers and to find broken ends, &c. Besides this, Mr. Walton urges that it is better to have four eyes watching the thread than two, as duvet and bouchon are in this way minimised.

In course of time, Messrs. Payen & Co. had to increase this filature by 300 bassines; but instead of increasing the steam-power reeling they fell back to hand-reeling, they also built several filatures in other places, and put in all of them hand-power turning.

Mr. Walton says that steam-power is enormously expensive, it takes 300 Rs. in France per bassine, which here would cost Rs. 400. To introduce steam-power into the whole of the filatures at Srinagar, he thinks it would cost more than Rs. 5,70,00; which would double his present outlay. He also says that so long as he has sufficient labour and not fully trained hands, he does not think it policy to introduce it; but later on, should labour get scarce and hands fully trained, it might be advisable to introduce it gradually. As an illustration; he says that by the end of June next he has to find 600 reelers for the 600 bassines about to be placed in the four new filatures. By having his present hand-power, he will be enabled to fill them with boys partially trained, whilst if he had not had hand-power turning he would have to train reelers who would take months to get to any reliable state of efficiency. The present cost of each reel-turner is only two annas per day. Steam-power with its wear and tear, extra supervision and maintenance, would be very costly in comparison. Let me now give my version. In Europe all the turning of the reels is done either with steam or water-power. I do not think steam-power is so costly as Mr. Walton thinks. when once it is established. There being no coal available at Srinagar is of course a great drawback; but if on calculation steam-power could not be economically employed, water-power could be applied and by being brought from the mountain streams at a sufficiently high level, a pressure sufficient could be found to turn the whole of the filatures without difficulty. Water-power in Italy is considerably used and it gives greater regularity than steam-power. Then there is the highly important question, I think, of education. The boys at present employed to turn the reels are young. Would it not be greatly better for their moral and future physical development for them to be at school with all its educational advantages and tone. I was very much pleased to see the superior intelligence of the boys in the Rev. C. E. Tyndale-Biscoe's school and I could not fail to think that there was something far better worth striving for in the well-being of a State than mere commercial economies. I should be glad to see power supplant hand-turning but only gradually, for as proficiency in reeling increases, a little more work might be done by applying power whether it be by steam or water. May I commend these ideas to the rulers of the State, and also the remarks on page 194 of my Report No. 1. In Europe, especially in England, Germany and France, great efforts are being made by successful manufacturers and others, to provide sanitary dwellings near their factories, thus greatly preventing epidemics, to see to the education of the young, to cultivate all that is worth having in a boy, trust, faithfulness and all that goes to make character and physique. I could mention many instances even in England where in this way the social position of the working classes is being elevated from what it was a few years ago, a very low state both intellectually and morally. Is it not worth considering in Kashmir?

No. 19.

Artificial Lighting of Reeling Filatures.

The question of artificial lighting is briefly treated of under this head at page 195 of my No. 1 Report. In considering this with Mr. Walton, the question of working hours and days naturally comes up. Mr. Walton argues, why should the filatures be worked in winter, when he can work them an equal time by closing during the winter months. He says he works on the Sundays. If he worked in the winter he would have to close on the Sundays. Then there are the Hindoo and Mahomedan winter holidays which would have to be deducted; he thinks there would be no advantage, and that he would not get as good quality of work done: the cost of warming the filatures and of the steam required would be a consideration, boilers and other repairs and furnaces rebuilt, etc. The whole of the reeling rooms would have to be glazed. Then there is no coal for gas-making, also the bad weather of winter and the unwillingness of the workpeople to come any long distance to work in winter time. Those living near are all employed in the winter in sorting cocoons and re-winding tests to the extent of one-third of the workpeople. During the summer months the whole of the salaried establishment is employed, and also as to the actual work-hours of the reelers as compared with those of Italy the following table will show how small is the difference.

The filatures of Srinagar are opened in spring on the first of March and close on the 15th of December, a little earlier or later according to the weather. They commence at 6 a.m. and work eleven hours per day, less half an hour for a meal at 10 a.m. They leave off work at 4-30 p.m.

Both Hindus and Mahomedans work on Sundays, the Italians and French do not.

Working months in Srinagar in the year.	Working months in Italy in the year.	Working days of 11 hours in Srinagar less half an hour for a meal.	Working days of 10 hours in Italy plus two hours for meals.	Srinagar working hours per day.	Working hours in Italy per day.
$9\frac{1}{2}$	10	290	305	$10\frac{1}{2}$	10

The working hours per year are shown by multiplying 290 by 10½ for Srinagar = 3045, and for Italy by multiplying 305 by 10 = 3050 which shows a gain of only five hours per year for Italy over Srinagar.

No. 20.

M. Arbousset's criticism of the Electric Light.

It is only necessary to refer to Report No. 1, page 195. It is indefensible.

No. 21.

Ventilation of Filatures.

Reference to my Report, No. 1, pages 195-6, will give explanation of a good and effective method.

No. 22.

Temperature of Cocoon-reeling and Batteuse.

The present system in Srinagar is the same as that in use in France and described in Report No. 1, page 196.

Mr. Walton informs me that the present Italian system in use in Srinagar is satisfactory and enables the reelers to reel more silk than by the French system.

No. 23.

Wages of Fileuses (reelers).

The Wages here contrast very favourably with those of France, the highest wages in Srinagar being more than three times cheaper. The four-skein or best reelers get 4 annas* per day, three-skein reelers $2\frac{3}{4}$ annas, two-skein reelers (the youngest ones) $2\frac{1}{4}$ annas, boys turning the reels 2 annas per day.

Thus in comparison.

For four-skein reelers, per day	4 annas.
One batteuse for two reelers ...	$1\frac{1}{4}$ „
Boy for turning	2 „
	$7\frac{1}{4}$ annas.

as against the French cost.

1 reeler in France	14 annas
Batteuse	5 „
Turning by power not costed	
	19 annas

Without reckoning the cost of turning, the difference in cost of labour is striking and points to a great advantage over the French, especially when the cost of steam-power is added.

No. 24.

Explanation of terms used in sericiculture and in the conditioning of silk at the Lyons Conditioning Houses.

I find these terms are well understood at Srinagar, and do not need further reference than to the list in the Report No. 1, pages 197-202.

It is a useful list and indicates how closely the Lyons Conditioning Houses watch and report

* An anna is equivalent to an English penny.

on quality of silk in all its different aspects, to tabulate the proportion of defectiveness and to show the purchaser what he is buying.

No. 25.

Quality of Silk to be aimed at in Kashmir.

Mr. Walton quite agrees with what is stated in Report No. 1, page 203, on this important head. It is an excellent guide as to the proper kind of output of Kashmir silk.

No. 26.

On M. Arbousset's recent expression to me of his opinion of Kashmir Reeling and Raw-silk of last year's production. (1902).

It is satisfactory and encouraging to learn from impartial outside authority that the reeling in Srinagar improves. This improvement will continue, as time goes on, in fact it is improving a little every day.

Mr. Walton is good enough to say that my visit has been a valuable stimulus both to the reelers and their overlookers, and has proved that by increased attention and care, a better quality of silk can be made.

With regard to M. Arbousset's advice as to the treatment of double cocoons, Report No. 1, page 204, of the above heading, it is well to give Mr. Walton's view. He satisfactorily shows me that it is better and more profitable here to reel them as far as possible and sell locally the inferior raw-silk produced for inferior purposes in the bazaar. I may mention that a double cocoon, called in Italy *doppio*, is a cocoon built up by two or three silkworms instead of one. This unity of interests at cocoon-making time produces a cocoon which cannot be reeled off from beginning to end like a cocoon constructed by a single silkworm, the threads being mixed together.

The silk produced from a maund of fresh cocoons when reeled to its utmost extent from double or doppio cocoons can be sold in the bazaars for Rs. 24. This is better than sending it to France, where only Rs. 20 can be got minus freight cost.

But this doppio yarn can be sold in London at 6s. per lb., which works out at Rs. 32 per maund of fresh cocoons.

In the bazaars here Mr. Walton can get 6 Rs. per seer at maund weight of 82 lbs. whereas in London it will fetch 8 Rs. per seer at maund weight of 75 lbs.

With regard to Mr. Walton's methods of output of the waste-silk, which includes waste (or what the French term déchets), doppio, ties, and nimtar, which latter is the refuse silk left on the chrysalides, they leave nothing to be desired.

No. 27.

White Cocoons and White Silk.

The production of white silk has recently

become an important European question, owing to a greatly increased demand for white raw-silk, and I have been much pressed both in England by manufacturers and in Lyons by the silk merchants there, to call the attention of Kashmir to this subject and to ask for supplies, which would be eagerly purchased.

I have discussed the subject with M. Arbousset, on my way to Kashmir, and also with Mr. Walton.

Unfortunately, white silk does not command a higher price in the markets than yellow, and I think it would be undesirable to change from yellow to white cocoon rearing for the following reasons, given to me both by M. Arbousset and Mr. Walton. Half a seer more silk per maund is obtainable from yellow cocoons than from white ones, this difference amounts to 10 Rs. more money from yellow than from white, and at 10 Rs. per maund on this year's Kashmir output, the loss would be two lacs of rupees, besides which white cocoons contain less gum and cause much duvet, and, in addition, the mulberry leaf as it is at present in Kashmir is hardly succulent enough for white cocoon rearing.

SUMMARY.

I would like to conclude my Report with a brief summary of the impressions and convictions I have received during my visit to Kashmir in respect of its State silk production.

I think the Maharaja and the State are highly to be congratulated on the unprecedentedly rapid growth of this important industry, and its sure promise of being a permanent one so long, as it seems to me, as it shall be steadily and carefully conducted by the State.

I cannot omit to put on record the excellent filature administration of Mr. Walton, the Director, and of the successful way in which he has taught the villagers over a very large area to become proficient breeders of silkworms and producers of cocoons.

It will be interesting to record the satisfactory progress of sericiculture during the last five or six years by repeating a few figures from the table at page 224.

In 1897 the industry was practically non-existent. It was in that year when, by direction of the Government of India, I made the first purchase of silkworm-eggs from Italy and France for Kashmir, amounting to £766. The purchases of eggs for the past six years have increased by leaps and bounds. For the requirements of the three years, 1901 to 1903, no less a sum has been expended in eggs than £12,999, or £4,333 per annum.

The number of persons engaged in attending to the rearing of silkworms has increased from 1,304 in 1898 to 44,245 in 1903, and what I think is especially satisfactory is the increased employment of Srinagar workpeople, mostly, I am informed, very poor and greatly needing work. In 1897 only about 100 persons were employed in filature work, whilst at the present time there are 3,500 persons busily engaged in the six filatures, and by June next this number will be increased to 5,000, on the completion of the four new additional filatures necessary to keep pace with the yearly increasing quantity of cocoons produced in the State.

All this could not have been accomplished without consummate knowledge, ability, and fixedness of purpose on the part of Mr. Walton, and the ready and unstinted help provided by the State. No small meed of acknowledgement must be accorded to my greatly esteemed friend, Lieut.-Col. Sir Adelbert C. Talbot, K.C.I.E., Resident in Kashmir during the first three years of the silk enterprise, for his foresight in urging forward the construction of filatures from the initial stage when the industry was but a problem, and also in the progressive interest he took in it, of which I have a vivid and very encouraging recollection. On his retirement in 1900, he had the satisfaction of knowing and reporting to the Government that the enterprise was a success, and a profitable one to the State.

Along with the name of Sir Adelbert Talbot must be mentioned that of Mr. Dane, whom I had the pleasure of meeting at the conference held at the India Office immediately before his departure from England to undertake the duties of Resident in Kashmir in 1901. The State owes much to his energy and administrative skill in putting the increased industry upon right working lines.

With regard to the filatures, I find them in excellent order and very fairly up to date. There are a few not very important exceptions which Mr. Walton is gradually having adjusted in accordance with the present practice in France, the particulars of which I have fully reported upon and discussed with him.

I have not avoided plainly pointing out defects, the most important of which is the present state of cocoon-reeling.

Without wishing in the least to cast any blame either on the filature assistants or on reelers, I feel bound to state with some disappointment that I am not satisfied with the present state of cocoon-reeling, which, although good on the whole is not equal to what I found in France on my way to Kashmir.

The reeling must be improved by the avoidance of nibbiness, duvet, bouchons and the better tying of knots, all of which are detracting from the market value of the silk to the extent of one to two shillings per lb. I must strongly urge increasing skill in the reelers, watchfulness and care in the turners, and above all unrelaxing vigilance of every assistant, both European and native.

But besides these suggestions, other points go

to make up the sum total of the requisite perfection attainable. The temperature of the water in the reeling bassines must be constantly minded as well as that of the cocoon preparing bassines. I advise the constant use of the thermometer; for an exact and invariable reeling temperature will go far to make the cocoons unwind more perfectly. The careful sorting of cocoons for reeling is also of considerable importance.

Besides these curative measures, I have suggested in this report the consideration at once of additional oversight assistance, and I repeat that whilst advocating a minimum of European assistants and a maximum of native assistants, it is for the State to consider carefully what maximum and minimum should be granted to Mr. Walton in order that he may have his filatures worked to the high state of perfection requisite to bring the silk up to European and Japanese standards.

If there are now only three European assistants to six filatures, they will be wholly inadequate to the continuous oversight of the greatly increased number of inexperienced reelers required for the four new filatures.

The total cost of the three present European assistants is Rs. 800 per month; additional assistants would begin at Rs. 150 per month each. When all the filatures are at work, one month's yield of silk of an advanced value of 1s. per lb., by better supervision, would in twenty days pay for this increase, even if there was one European to each filature.

Considering the wish of the State to prefer native employées, Mr. Walton tells me he could manage with one European to each two filatures, which would be an increase of two when the four new filatures are finished. On this point I do not agree with him, for I consider, judging from my Continental experience, that one European assistant to each filature is an absolute necessity to bring up the silk to the standard, and better still with two.

At the times of distributing the eggs to the villages the European assistants must of necessity be away from their filature oversight, as also when the cocoons have to be weighed in from the country—a large undertaking.

I contend that each filature should have at least one European assistant never absent. Incessant vigilance is the only preventive of careless reeling.

I would suggest that it would be a good thing to send down four or six elderly boys from England to the South of France simply to learn all about cocoon-reeling and to be instructed at one or more filatures there. I could easily obtain the necessary permission.

There would be no necessity for any lengthened stay, as a knowledge of cocoon-reeling only would be required, and not such a study of sericiculture generally as is given in the Government Schools of France and Italy. If a promise of the State were given to parents of boys of something like a permanent prospect of employment, they would not grudge the expense of sending their sons to the French filatures.

It is of especial importance now the Industry has assumed such large proportions, that drastic and sudden changes or alterations should be avoided, and that any suggestions of improvement from whomsoever they may arise, should be first tried experimentally. Even as to what I have written about turning the filature reels by steam or water-power, if it should ever be thought desirable to adopt it, trials only should be made in one of the filatures and their results carefully examined and compared after working for some time and before any large outlay is incurred.

The sericicultural ship is now in deep water and requires to be steered and worked with great steadiness and by the avoidance of disturbing elements.

In closing this report, which has been to me a labour of love, I hope on my return home to do what I can to institute preliminaries for the successful introduction of silk-weaving on the lines I ventured to submit to His Highness the Maharaja, General Raja Sir Amar Singh, the Resident and the Durbar, who have so kindly received me, and who fully concur in this suggestion.

That the Silk Industry in which I for so long have taken a purely disinterested interest may continue to flourish and be a comfort to His Highness the Maharaja and a blessing to the people of his State is the earnest prayer of

Their humble Servant,

THOMAS WARDLE.

CHAPTER XVI.

THIRD REPORT ON KASHMIR SERICICULTURE, WRITTEN AT LEEK DURING THE LATTER HALF OF 1903.

Leek,

January, 1904.

To the Under Secretary of State for India.

Sir,

I now have the pleasure of completing my final Report, the result of discussion and inquiries in France and Italy, concerning a few points where more technical knowledge was desirable, and upon one or two divergences of opinion between Mr. Walton and myself, which M. Arbousset, Mr. Walton and myself had already discussed to some extent in my Report No. 2.

I have endeavoured to accentuate the results of these enquiries and to put them in the clearest form.

Mulberry Cultivation.

First, as to the cultivation of the Mulberry, the practice of which in Kashmir was so bad as to have caused me very unwillingly to pass some severe strictures, as shown on pages 209-213 of my No. 2 Report.

Following the practice of France and Italy, the Mulberry-tree should be pruned as little as possible, about once in four or five years to shorten the branches, this should always be done in the winter, preferably in February, and before the leaves appear.

In the Cevennes the pruning used to be done annually, but it was found to cause the trees to perish, and for that reason the practice has been given up. The leaf that grows from a branch two to four years old, is far better for the silkworm and more nourishing than that which grows on a branch of the same year.

When a tree is pruned in February its leaves must not be picked in Spring, but only in the following year.

I have already described in a previous letter, the method of gathering the leaves, but it is better that I add the following remarks:

The branches must never be cut in order to get the leaves off.

The leaves must be picked off by hand and put in a bag tied round the waist, and if the trees are very high a bag kept open by a circular wooden ring, must be used, and one that can be hung on the trees with a wooden hook or pole.

I sent a drawing of the apparatus to Mr. Colvin

last August, and young Mr. Thomas will bring an example of it as used in the Cevennes, and he will also have been instructed in its use.

Ventilation of Magnaneries.

The Magnaneries must be ventilated, but Mr. Walton's opinion I find to be correct, that variations of temperature from 68° F. must be avoided.

When the weather is cold, *a fire must be lighted*, the draught caused by the fire will give sufficient airing.

On the contrary, when the weather is hot, it is necessary greatly to air the magnaneries without fear of draughts, but the openings must be closed if the temperature gets cooler or becomes damp.

Etouffoir versus Séchoir.

I have had a prolonged discussion with M. Arbousset as to the relative merits of each. He maintains his former opinion, and prefers the Etouffoir.

With the Etouffoir the cocoons cannot be burned, and he repeats that the cocoons etouffé by the Séchoir unwind better during the three or four of the first months after the gathering of the cocoons, and give a silk which has not so much fluff, whilst the cocoons etouffé with the Etouffoir unwind better than the Séchoired ones during the rest of the year, their grès, or gum, being less dry; but he admits that in damp climates like

China, the Séchoir is preferable because with it cocoons can be dried and kept. He thinks on learning that the Kashmir climate is damp, it may be better to use the Séchoir, as Mr. Walton does, and he quite approves of his only half drying the cocoons; in that way the grès, or gum, does not get so hard, and the cocoon is more easily unwound. In the Cevennes during the months from June to September, the average of dampness registered by the polymetre is from 30 to 40 degrees. It would be interesting if Mr. Walton would daily register the hygrometric state of the climate by means of his polymetre, and to furnish us with the records.

Cocoon-reeling Supervision.

The reeling supervision is greater in France and Italy than I found it in Kashmir. In the reeling filatures in Alais, to each 60 bassines there is one male director, two skilled women superintendents, and one tester, one of the principals generally superintending the whole work.

At St. Cristol for 40 bassines there is one director, and two women superintendents, one of which is a tester.

All this confirms the opinion I so strongly expressed when in Srinagar, and also on pages 227-8 and 243-4 of my No. 2 Report, that the filatures lack more European and native superintendence, and I fear that only by resorting to European procedure will the reeling be brought up to the high mark of the French and Italian qualities.

It is necessary to do all that is possible to avoid defects in the silk thread, but it is useless to try and satisfy every buyer, who will always try to find defects in order to obtain the silk at a reduced rate.

Motive Power.

In spite of Mr. Walton's good reasons, I retain firmly my belief and opinion that it is preferable for the regularity of the motion, the facility of un-reeling the cocoons, and for the regularity of the titre, to have a mechanical motor, either hydraulic, or steam-power, or by electric motors, if this latter system can be economically installed.

My opinion is fully endorsed by that of several of the best European reelers, besides which, all of them practise it.

The expense, even with a steam-engine, would be much less than Mr. Walton thinks; in fact much less than with boys working well even at two annas per day.

If it was not necessary to have steam to heat the pans, and that it was necessary to instal boilers and steam-engines, the expense would be considerable, but as long as there are already boilers to give steam to heat the pans, the extra expense of firing for one steam-engine would be insignificant.

One hundred boy turners at two annas per day cost 16s. 8d., and one three-horse power steam-engine would turn 100 reels at an extra cost of less than one anna per day with coal, but with wood the cost would be somewhat higher.

A water-motor, being even more regular than steam-power, would be preferable, and would cost less when once applied.

Cocoon-reeling and Apprenticeship.

M. Arbousset has sent Mr. Walton a photograph of one of the new machines which have now begun to be used in the French filatures. It is constructed with a separate spindle for each skein of silk, and is put out of motion by a lever above the head of the reeler. The advantage in this is that the reeler is able to continue her work at the same time the girl joins the broken ends.

With this new machine the apprenticeship or probation would be made quite as well as with the boy turners, and even better.

Between each two pans there is a small pan, called a "petite bassine batteuse," on a bank or tray which runs the length of the filature. This is the little bassine which is placed in front of and between two fileuses; it is the mode employed in Italy for the battage. The battage in Italy is not done in the same way as in France, there is an apprentice who places a handful of cocoons in the little bassine, and beats them with the brush, which is much smaller than the one employed in France, only being about half the size. She beats the new cocoons and the detached cocoons separately, and passes them to the reeler to clean as soon as the threads have caught her small heather brush, or rice-straw brush. It is these little girls who working all day under the eyes of the reelers, try to make a good and short apprenticeship.

There are also the young girls who tie the knots or join the ends together again, one for every five workers, and who are placed between the worker and the machines; they tie together the broken threads; a saving of time and a regularity is thus effected. They are able, when tying the knots, to obtain a useful and growing experience with the reeling, and therefore they soon learn how to reel cocoons.

This shows clearly that even if the staff were smaller, those undergoing apprenticeship or probation would be just as successful, and besides the work would be more quickly and better done, especially if a mechanical motor were substituted for boy turners.

M. Arbousset thinks that it is impossible for really good silk to be produced without the use of a motor for turning the spindles, as a young boy is unable to turn with the regularity necessary, and unsteady turning cannot produce good silk.

Vice Rédhibitoire.

I have fully described this defect on page 202 of my No. 1 Report.

In consequence of an apparently well-founded charge, as well as a large claim for raw-silk said to be damaged by dermestes or mites made by the eminent firm of Messrs. Chabrières, Morel et Cie., of Lyons, I discussed the matter with Mr. Walton at Srinagar, and also carefully examined much of the raw-silk about to be sent to England, but we were unable to find any trace of the defect, or any trace of insect life, exuviæ, microscopic or otherwise, likely to gnaw, eat, or damage the silk.

On my return from Kashmir, in May, 1903, I called on Messrs. Chabrières, Morel et Cie., at Lyons, to further investigate the defect, and to fully discuss it with them. I was assured that the cause was what they had stated; and that in their long experience as merchants of raw-silk they had had to make many claims from this cause; they affirmed there could be no doubt about it, and that it was most important for the future of Kashmir silk that a remedy for it should be found.

I helped to open and examine several bales, which at my request were brought up from their stores to their office in Lafont, Lyons.

I found ample evidence of defective silk, and instructed them to open all the bales, about sixty in number, which they then held in stock, and to send all defective portions to Messrs. Durant, Bevan & Co., to be forwarded to me for a minute investigation, for I was not then convinced that the cause was what they so confidently stated, or "moth-eaten" according to their ordinary parlance.

Messrs. Durant, Bevan & Co., subsequently sent me all the defective hanks they received from Chabrières, Morel et Cie., amounting to nearly 100 lbs.

I gave every hank a thorough examination, chiefly microscopic.

This occupied me upwards of a week of close work. I am glad to say I found no trace of animal or insect life, microscopic or otherwise, but there was ample evidence of damage of another kind.

In many places, chiefly at the end of the hanks which had naturally been the most subject to friction, the silk fibres were much cut and frayed, and so rubbed that the silk thread itself was only recognisable by aid of the microscope. It was also discoloured, and gave rather the appearance of dermestes exuviæ. I came to the conclusion that the fault lay with the carrier and not with any entomological destructiveness, but to make sure of this I sent some of the defective hanks to the highest authorities in Milan, namely, the "Société Anonyme Coopérative avec capital illimité pour le conditionnement et l'essai des soies et autres textiles à Milan, Siège Principal, Via Ciovasso 11," who have issued a most interesting brochure entitled, Compte-rendu des recherches faites de 1894 à 1899, Laboratoire d'Etudes de la Soie. Published in Milan by L. Zanaboni & Gabuzzi, 12 Via Fontana, 1900. The reply I received was as follows:

"There is nothing in the appearance of the silk which will authorise us to declare that the silk has been cut by dermestes or other insects harmful to silk. Our opinion is that the defect has been caused during the transport, in which sufficient care has not been employed, either by the rubbing of the bales against each other, or against the cords which tie the bales."

But to prevent my assigning an opinion which might yet be found controvertible, I wrote also to the Director of the Royal Bacological Station, Padua, and append a copy of his reply.

Padua,

3rd August, 1903.

R. Stazione Bacologica Sperimentale.

Dear Sir,

I have given your silk a long examination, and I am of opinion after many experiments that the so-called "moth-eaten" defect depends exclusively upon a too strong ligature, or from some iron instrument.

I have found no trace of insects, and it is impossible that an insect could cause the mischief.

Yours very truly,

E. Quajat.

There is therefore now no doubt that the subject has been thoroughly thrashed out, and whilst the rédhibitoire idea could not be applied legally or justly in this case, I have recommended Messrs. Durant, Bevan & Co., to make a satisfactory allowance to Chabrières, Morel et Cie, and settle the matter. This they have done, and have sold the damaged portions at their lowered value.

Although the silk has been saved from a bad reputation, and from a liability to claims more or less serious for defects *always made the most of*, it is absolutely necessary that better arrangements should be made for conveying the bales from Srinagar to London, as much as possible friction-proof.

I think the damage is caused in transit between Srinagar and Rawal Pindi, along the steep gradients which involve much jolting, and possibly by too tight cording; anyhow the defect is preventable, and it is of interest to find that most of the silk sent to London has not been complained of, in fact, Messrs. Chabrières, Morel et Cie., are the only firm who have found fault or claimed damage.

The Sorting and Choice of Cocoons for Sale.

On this question of high importance the opinions I have gathered go all in one way.

The samples of cocoons that were sent to me, as well as those I took to M. Arbousset, were not sufficiently well chosen to be sold on the Marseilles market.

In reeling the good cocoons, the poor ones must not be mixed with them, for in the battage of cocoons, the batteuse brush loosens and destroys the poor cocoons, and the good ones do not become sufficiently boiled or prepared.

The silk from poor cocoons and those "satinés" is also more fluffy than the silk from good cocoons. For this reason it is necessary that the cocoons should be carefully selected before they are sent to the filature or sent for sale. The poor cocoons must be reeled separately for the purpose of making grège or raw-silk of the second quality, which is sold generally at four to five francs per kilo., or from 1s. 6d. to 2s. per lb. less than the first quality.

M. Arbousset endorses my opinion that one European superintendent is necessary for each filature, and hopes that the experiment that is going to be tried with Mr. Thomas will succeed and will favour our joint opinions.

When I left Srinagar, the mulberry trees, owing to the inclemency of the weather, were about three weeks late. I was then very apprehensive of the success of the silkworm crop, and I am sorry now to learn from the authorities in Srinagar, that owing to a heat-wave or as it is called in French "la touffe" coming on suddenly and severely before the worms were old enough to stand it, there has been a loss of about 25 per cent. of silkworms. On learning this I at once wrote to Mr. Colvin and strongly recommended that in future from 20 to 25 per cent. more eggs should be purchased and distributed than at present, as the ten filatures ought to be fully kept at work, and any surplusage of cocoons could very easily be sold at a profit in Marseilles.

With regard to the eggs purchased last year for this year's crop, I would like to suggest that the hatching and progress of the rearing should be very carefully observed, because owing to great shortness in France of silkworm-eggs there has been the greatest difficulty in supplying the requisite quantity, and although the eggs are always guaranteed each year to me by M. Arbousset, in whom I have the fullest confidence, to be free from pebrine, it would be as well to see if any disease creeps in and for extra precautions to be exercised.

Within the last few days I am informed in a letter I have received from Major Kaye, of his approval of my suggestion to order more seed than is required, and so guard against having to shut down the factories if there should be a short crop of cocoons.

I am glad to be informed too by him that 25 per cent. more country people have volunteered to apply for seed next season, making up the number to 13,000, and as the seed distributed is one ounce per family of four or five, this will bring up the people employed in this industry, young and old, to between 52,000 and 65,000, a most pleasing industrial feature, for they are paid handsomely for all cocoons they bring in.

Major Kaye also informs me that the newly-employed female reelers are on the whole doing well, many of them reeling four skeins of good silk, and he believes the experiment has all the elements of success in it.

I am glad to be informed by him that the weaving industry should be started and continued as a State branch.

Now that the discussion and correspondence on points raised in Srinagar on my No. 2 Report are completed, M. Arbousset modestly desires me to ask that his name may be allowed to be mentioned amongst others, as having assisted in the establishment of sericiculture in Kashmir, as much by the carefully chosen eggs he has all along sent, as by the advice and opinions he has so unstintingly given.

With regard to this, I have no hesitation in saying that Kashmir sericiculture is under a deep debt of gratitude to him for being so reliable in the great trust reposed in him, as well as for the advice he has always been ready to give, and for his generosity in training young Mr. Thomas in his filatures, and in obtaining an introduction for him to work in the filature of another reeler of Cevennes reputation.

I wish to state that the 10 lbs. of 24 deniers raw-silk reeled from white cocoons which was sent to me by Mr. Walton on 24th November last for weaving experiments, is turning out most satisfactorily, both in regularity of cloth and in the improved augmentation of size over and above the $\frac{9}{11}$ or $\frac{11}{13}$ deniers.

I hope soon to send specimens of the woven fabric and its results in dyeing and finishing.

Suggestion for Improvements in the Tying up of Frisons or Knubbs for the Market.

The present plan of fastening the bunches of waste or knubbs with tightly twisted bands involves a great and needless waste of labour in untwisting the aforesaid bands, and untying the knots, and is greatly objected to.

A more loose mode of tying the bunches, and in much larger bunches, would be much preferred and would seem not to be difficult to do.

I again called at Messrs. Durant, Bevan & Co. a short time ago and examined a number of bales they had in their stores, and found a good many of them were very much rubbed and abraded by imperfect carriage, some of them being in holes close to the ends of the bales.

New French Reeling Modification.

This recent improvement prevents the stoppage of the reeling of all the four skeins when the end of one of them breaks and enables the reeler to proceed with the reeling of the other three skeins not broken, so that when one end breaks, only one skein stops.

The silk thread is carried from the bassine over the tavellettes and the two portesbouts to the reel or "dévidoir" in exactly the same way as with the ordinary reel or "dévidoir" but instead of there being one long reel holding four skeins, the reel or "dévidoir" is divided into four narrower ones.

A photograph has been sent to Mr. Walton, and when Mr. Thomas arrives he will be able to point out the details of the new "dévidoir" and the connections for stopping and starting each one. I have asked him to take out with him the important parts of the mechanism.

This improvement enables a boy for every five reelers, to be employed in the filature, his work is simply to piece up the broken or detached ends, whilst the reeler continues laying on the cocoon threads as often as required. The boy works behind the reelers.

I have been in close touch, by correspondence, with Mr. Thomas during his stay at Alais. I am sure his work there will prove very useful to Kashmir. He is fully conversant with all the details of the filature and subsidiary work carried on in the Cevennes. He has sent me many details of procedure there of greater or less importance concerning cocoons, reeling, battage, evenness of thread, avénage, water temperatures, overlooking, duties of managers in France, elasticity and strength, duvet and bouchon, water preparation, labour particulars, titrage of cocoons, regularity of thread, progression of sizes, various "essais," lances, croissures, tying of knots, porte-filières, &c., but as he has them all carefully written down, I need not add them to my report; it may however be desirable to print them in Srinagar as a record of his most useful study of economical French procedure in filature work.

Durant's New Offices.

I have been informed by Messrs. Durant, Bevan & Co. that they have within the last fortnight, opened a bureau and warehouse in Lyons which will give them greater and closer facilities for the exploiting of Kashmir silk amongst the silk manufacturers of France than they have hitherto had. Mr. Bevan is there at the present time installing a responsible representative of the firm, and making the necessary arrangements for future prosecution of their business as Brokers of raw-silk in France.

As my life is drawing near its close, I am I hope

not unnaturally desirous of putting on record some statement of my life work for India: and this is my apology for recapitulating thus fully my deep personal and philanthropic interest in the welfare of India.

To me my work for India has been indeed a labour of love, which has increased with years, and still continues.

I have never had any desire to work for the silk interests of India from pecuniary motives, and although the time spent has been enormous, and the hobby a costly one, I have no regrets; on the contrary, I cherish a pride in my efforts, I hope justified by the success which has attended them, and by their potentialities so far as silk, sericiculture and silk weaving are concerned in the future of India.

Thus an historical Industry which had been allowed absolutely to die out after repeated attempts from lack of knowledge and scientific procedure, and which employed only a few hands in 1896-7, now gives employment to more than 50,000 men, women and children, all natives of the country, their number increasing rapidly each succeeding year; the silkworm rearers nearly all working in their own homes in the villages of the valley of Kashmir, and the reelers working in the factories or filatures in Srinagar, the only Europeans engaged in this colossal industry being the Directing Superintendent and his five assistants.

This is a dramatic result, and fortunately as sound and sure as it is startling. And it is not more gratifying in itself than in the solid hope it justifies of similar beneficent results being obtained in the revival, and even the introduction of other industries into India, of the ruri-urban type of sericiculture and silk reeling and weaving.

I beg to offer this Report to His Majesty's Government of India, acknowledging with my warmest thanks the kind encouragement I have invariably received.

I have the honour to remain,

Yours faithfully,

THOMAS WARDLE.

CHAPTER XVII.

THE ROYAL BACOLOGICAL STATION, PADUA, ITALY.

I HAVE lately received from Professor Quajat, the Assistant-Director of the Royal Sericicultural Laboratory at Padua, a most interesting brochure containing an account of the curriculum at the Royal Bacological Station at Padua, and the scientific sericicultural investigations regularly carried on there.

In giving a few quotations from this brochure about the Institution and its valuable educational and scientific work, it will serve to emphasise in a marked way my former, as well as present, strong convictions and recommendations to the Government of India on the importance of having in India a similar Imperial sericicultural educational and scientific establishment, and all the more so that the Government of Italy not being content with this one famous agricultural teaching institution in Padua, has actually 59 sericicultural observatories and laboratories now working in the Kingdom of Italy under the direction and dependence of the Station at Padua. The names of these stations are given in this brochure, which is entitled "The Royal Sericicultural Laboratory of Padua, at the Universal Exhibition of St. Louis, being a summary of information of the functions of this important Central Establishment at Padua, 1904."

It is proved that these sericicultural laboratories have given mighty impulse to the revival of the silk industry in those extensive areas in Italy where the inhabitants mostly apply themselves to its production.

The brochure states that the diffusion of healthy eggs incessantly brings further benefit to the industry, especially when it is known that they have been produced under the scientific direction of the Government Establishment.

Many breeders of silkworms, Italian and foreign, come personally to discuss the difficulties that happen to them, and to be more closely acquainted with the establishment in all its details.

Besides the curriculum work, one of the duties of the laboratory obliges it to execute microscopic analyses of eggs, chrysalides and moths, at the request of private applicants, on payment of a fee.

The Padua Laboratory is under the direction of my esteemed friends the learned Cavaliere E. Verson and of his learned colleague, Cavaliere E. Quajat, the Assistant Director, a list of whose Sericicultural Memoirs is to be found in the brochure, occupying no less than sixteen pages.

This laboratory was founded by Government in April, 1871, and has already existed for 33 years.

The Ministry of Agriculture, supported by the Province, the Commune and the Agricultural Association of Encouragement of the local industries, provides for its maintenance, and all these corporations have their representatives in the Board of Administration.

The persons appertaining to the Institution, are:—a Director, who directs and surveys the work, gives impulse to scientific researches and is charged with the regularity of the different administrative operations; a Vice-Director, who assists the former and substitutes him when occasion requires; an Assistant; a Secretary, and persons of common service.

The building occupied by the Laboratory belongs to the province.

Adjoining the establishment is a certain space of ground (about a hectare and a half) cultivated with mulberry trees, of which there is a collection of more than 30 different kinds. In the middle of the grounds rises the silkworm nursery; it is completely isolated, and annexed to it is a hothouse for the production of the leaf for the early instruction.

The Cabinets are provided with all that is required for the sericicultural researches. Microscopes, balances of precision, physical and chemical apparatus, sterilizing and cultivation apparatus, are the scientific instruments with which the laboratories are provided. On the other hand, great care has been taken in collecting day by day all that is connected with the technical practice of the Silk Industry. And thus the laboratory now possesses a rich museum, containing numerous models of hibernation and incubation machines; all kinds of implements and apparatus for rearing silkworms, for isolating and selecting moths and cocoons, for suffocating the latter, and in short for all the operations necessary in the production of silkworm-eggs.

The most varied races of silkworms are represented in more than 1,600 vases, containing cocoons and raw-silks. Silk-winders, reels, serimetres and serigraphs and several kinds of basins are used for analyzing the threads of each cocoon, as well as the raw-silk formed by several threads.

Plaster models of healthy and unhealthy worms and large coloured tablets, show in all its details, the anatomy of the Bombyx mori; its different evolutive phases, and the delicate structure of its tissues may be seen by the observer in a collection of 8,000 objective laminae bearing more than 70,000 microscopic sections.

I have had the honour of contributing from time to time important sericicultural materielle to this museum for which I have received the thanks of the Italian Government and an offer of an Order of Decoration, but which I was not permitted by rules of the British Government to receive,

The Library contains more than 2000 volumes, representing together with technical and scientific works and magazines an average value of 20,000 francs.

The traditional empiricism that governed not long ago the most important treatment for the proper preservation of eggs, has yielded to a rational method, strictly agreeing with the exigencies of a natural evolution. The nature of the diseases that unfortunately still afflict the silkworm, has been studied in all its manifestations, to facilitate the knowledge and application of the best means to struggle against them. Systems of disinfecting the nurseries have been settled, without however damaging the products of the rearing, and several other important problems that greatly interest sericicultural industry, have had enlightening.

For sericicultural details, the reader may consult the 31 volumes of periodical publications, edited by the Establishment, and the considerable number of Memoirs it publishes in scientific papers or in the reports delivered at several scientific Academies.

This Sericicultural Establishment has not only greatly contributed, with its publications, to render silkworm-breeders more careful and watchful in the practice of this industry, but it possesses two other means to reach the same aim more efficaciously; the courses of teaching and the indefatigable concurrence of sericicultural laboratories. The usual teaching is imparted in two separate courses; the first for men and the second for women, and regulated so that they may reach practically useful results, they take up all the activity of the persons employed at the Establishment for more than four months every year. But this heavy task is widely rewarded by the kind reception of the public, if it is permitted to judge by the number of those who draw advantage from their labour. And a considerable advantage is indeed drawn from it, even materially speaking notwithstanding the sad vicissitudes to which the Silk Industry has been subjected during the last ten years.

It is gratifying to find that of 909 pupils who have passed through this Institution, 500 at least have applied themselves to sericiculture as their chief, if not their only, means of living.

Two important publications have recently been issued by the Director and Vice-Director of the R. Stazione Bacologica, the titles and particulars of which are given at the end of chapter xviii.

CHAPTER XVIII.

SILK AND THE SILKWORM: PRACTICAL TREATISES ON SERICICULTURE.

ONE of the most useful and reliable aids to Sericiculture is Monsieur Laurent de l'Arbousset's book entitled "Cours de la Sericulture Pratique," published by C. Castagnier, Alais (Gard) France. It is a book I earnestly recommend to all persons interested in the subject.

My daughter, Elizabeth Wardle, has translated it into English, and I have edited it. It will be published shortly by Mr. Eaton, Leek, Staffordshire and Messrs. Simpkin, Marshall & Co., Ltd., Paternoster Row, London, and will be found most useful both in India, our Colonies and America.

My friend, Mr. Nitya Gopal Mukerji, of Naya Dumka, Southal Paraganas, Bengal, India, who was Sericiculturally educated at Padua and Montpellier, the Italian and French Sericicultural Government Stations, and also under the late Professor Pasteur at Paris, has also written an excellent treatise on the same subject especially adapted to Sericiculture in India. It is entitled: "Handbook of Sericiculture," and is published by the Bengal Secretariat Press, Calcutta, both in Hindustani and in English.

The following are two important publications issued from the R. Stazione Bacologica :—

1. Il Filugello e l'Arte Sericola, Trattato Teorico-pratico illustrato da 85 incisioni intercalate nel testo, by E. Verson ed E. Quajat, Direttore e Vice-Direttore. Fratelli Drucker, Padova. 1896. Price L.10.
2. Dei Bozzoli, by E. Dott. Quajat, Vice-Direttore della R. Stazione Bacologica di Padova, Più pregevoli che preparano i lepidotteri setiferi, con 50 tavole contenenti oltre 140 clichés. Padova, Fratelli Drucker, 1904. Price L.10.

CHAPTER XIX.

JOURNEY TO ALAIS. BOMBAY. BOMBAY TO SRINAGAR, VIA RAWAL PINDI. SILK FILATURES, ETC. SERICICULTURE. MY REASONS FOR GOING TO KASHMIR.

I VENTURE, I hope, with some appropriateness, to include in this book, two letters I wrote to the Leek newspapers during my absence from England, one on the Sericicultural operations in Kashmir, the second a description of its Geology, Natural History, and Sport, also with registers of Temperatures and varying Barometric pressures during the journey.

From the Leek Post, Saturday May 9th, 1903.

We have received the following letter from Sir Thomas Wardle, describing his journey from Bombay to the Vale of Kashmir, and to the Silk filatures in Srinagar which he has been the means of establishing during the past few years, an account of which will be of the utmost interest to our readers :—

TO THE EDITOR OF THE LEEK POST.

Dear Sir.—On my way out to India I stayed a

PLATE 9.

One of the two State Elephants of H.H. the Maharaja Sahib of Jammu and Kashmir, as caparisoned for the Delhi Durbar of 1902.

week at Lyons and the further south of France, to get the opinions of silk manufacturers and merchants on Kashmir silk, and to study up-to-date procedure of cocoon-reeling, etc.

By the kindness of two of the best filature proprietors in the Cevennes at Alais, I was permitted to work at cocoon-reeling. I made a lengthy report, which took up much of the time during the voyage to compile, both from notes and memory.

I journeyed alone from Bombay, on Monday, the 9th March, by express night and day, passing Bhopal, Gwalior, Agra, Delhi, Amritsar, Lahore and Rawal Pindi on the way, arriving at Sialkot on the following Thursday, where I met by appointment Mr. Colvin, the Resident of Kashmir. After a day's discussion with him, I went on to Jammu, where I met the Maharaja of Kashmir and Jammu, his brother, General Raja Sir Amar Singh, and the members of the Durbar in Council His Highness the Maharaja warmly thanked me for the interest I was taking in Kashmir Sericiculture, and after hearing my report, and what I had to say, in an interview lasting nearly two hours, he ordered my report to be printed, and after a most cordial reception at the Palace, his brother took me to his house adjoining the Palace. I stayed several days at the Residency, having an elephant and also a carriage placed at my disposal (plate 9). I saw the great annual Hindoo festival of Holi; waited until the Resident came up from the Sialkot Residency, and then, after much discussion with Sir Amar Singh and the Resident, I left for Rawal Pindi, en route for Srinagar, the capital of Kashmir.

When I arrived at Rawal Pindi, the nearest railway station to Kashmir, 200 miles away, I was informed that the road up to the Kashmir Valley was blocked with snow, and would not be open for ten days; that part of the journey having to be done by Tonga, occupying four days, resting at Dak bungalows each night. I went on to Peshawar for a few days, waiting for the snow to melt. Here the weather was almost tropical. I went a long drive up the Khyber Pass. I returned to Rawal Pindi after four days stay, and learned I might risk the journey if I walked through ten miles of snow. I waited two more days, and news came by wire that the snow was being cleared and leaving only a walk through it of two miles, at Murree, 7,200 feet above the sea. I decided to venture, and set out alone in a tonga and pair, changing horses every five or six miles. The first stage was over the plain of Tret, 25 miles; the second to Dhulia, 47 miles from Rawal Pindi; leaving Dhulia next morning at 7 o'clock, I drove on to Baramulla, 102 miles, arriving there at 9 p.m. the same day, having had a snowy, cold, and very dangerous journey, under high cliffs and over precipitous heights most of the way. My tonga was stopped at one place by a large fall of cliff which had just smashed an ecca laden with luggage. I had to jump out of the tonga to avoid another fall of rock. The

PLATE 29.

PLATE 33

Srinagar Silk Filature. Length 435 feet, width 42 feet. Containing 550 workers.

PLATE 34.

Interior of Filature in operation.

river Jhelum was foaming in flood 1,000 feet below.

Baramulla is the western entrance to the Vale of Kashmir, and only 30 miles from Srinagar, 5,235 feet above the sea (plate 29). The valley is surrounded by snow-clad mountains of great grandeur, some of them from 12,000 to 26,000 feet high. The valley is chiefly flat, and is the bed of an ancient lake.

I arrived at Srinagar at mid-day on the fourth day. The Governor of Kashmir called on me soon after my arrival, and said his instructions were to see that I lacked nothing, and that as a State guest I should receive every attention to my comfort. His name is Pundit Man Mohan Nath Kaul Sahib; he was most polite. In the afternoon, Mr. Walton, the able Director of Sericiculture here, drove me to the silk filatures. There were six fine factories or filatures for reeling and storing cocoons; each filature is 435 feet long, by 42 feet wide, and has two rows of reeling bassines (see plates 33, 34, 35, 36, 37). There are 550 men and boys in each filature, a sight perhaps unequalled in Europe. Four more filatures are in course of erection, and will be completed before June, when the new crop of cocoons is ready; but about two-thirds of last year's cocoons are waiting to be reeled, owing to the large crop and inadequate reeling machinery. To think this immense industry had all developed from the £600 to £700 worth of silkworm-eggs I bought six years ago seemed wonderful, and made one feel thankful.

The number of workpeople now employed is as follows:—Eggs distributed to 11,150 families; Mr. Walton informs me that the quantity of eggs given out depends on the size of the families, and that an average of four per family who partake in the rearing is a correct estimate; so multiplying 11,150 by 4, there were in 1903 the following:—

Silkworm rearers	44,600
Reelers and Turners in the present six filatures	3,300
Sorters	200
Packers and deniering	60
Firemen, wood-cutters (there is no coal) ..	50
Carting wood	40
	48,250
Additional reelers to be employed in June in the new filatures	2,200
	50,450

In 1898 the eggs I purchased in 1897 were distributed amongst 500 heads of families in the villages; increase has gone on each year until at the present time £4,000 worth of eggs are supplied to no less than 11,150 villagers; they occupy their families and relations to the extent of 44,600, whilst the number of reelers and turners has increased in the last five years from 80 to 3,300. This reads like a fairy tale; but it is accurate and true; thus making up a total of almost 50,000 people, old and young, in an industry, which, six years ago, did not exist.

PLATE 35.

View showing the ends of Six of the Ten Cocoon-Reeling Filatures at Srinagar.

PLATE 36.

Baleing Room of the Srinagar Filatures.

PLATE 37.

Two Hibernation Houses for Silk-worm Eggs.

The Chief Judge of Kashmir, Mr. Mukerji, and Dr. Mitra Rai Bahadur, State Medical Officer, told me yesterday that had it not been for the successful introduction of Sericiculture, Srinagar would have been famine stricken, and that even now a very large number of people in the city are all but starving, and needing employment.

I have suggested a sister industry, that of silk weaving, which, by the exceedingly clever fingers of the Kashmiris, can, I feel sure, be successfully worked, and I have instructions from the Maharaja, his brother, and the Resident, to introduce it.

It has great possibilities, and may be as important in time as that of Sericiculture.

But to return; I have driven through the length of this valley of wondrous beauty, about 80 miles, and 35 miles wide. It seems full of mulberry trees, and secures the home of an industry, which even after its present marvellous expansion is yet but the nucleus of an immense future development, if it can be retained and managed by the State, and so preserved from commercial rivalries.

My mission here has several aspects. First, to show the great advantage to the State to retain the industry and to keep it free from private enterprise and company promotion; second, to provide employment for a large population greatly in need of it and to help them to be fitted for it.

I have never seen a people so marvellously clever with their fingers as the Kashmiris. It is a sight to go down the bassines in the filatures and see the beautiful way the men and boys can handle the fine threads of the cocoons, even those who have been engaged in it for a few weeks only. They do it with a readiness and skill which compare favourably with the south of Europe fileuses, who have gone through an apprenticeship to learn cocoon-reeling.

Thirdly, another object I am working for is to help to perfect the raw-silk, so that it may hold its own against that of Italy and France.

It nearly does that at present. Already improvements are progressing, and I hope when the mulberry cultivation is better attended to, to see my aims accomplished, and at no distant date, and that this industry, already very profitable to the State, will have a most prosperous future.

Fourth, to obtain a reliable balance-sheet; fifth to suggest silk weaving.

In a second letter, I will describe what I have seen of the geology, mountain sport, and natural history of Kashmir, from the ubiquitous sparrow to the imperial eagle and lordly barasingh.

I wish the weather were warmer; it is very cold and chilly.

Yours very truly,

THOMAS WARDLE.

CHAPTER XX.

KASHMIR MOUNTAIN ALTITUDES, GEOLOGY, FOSSILS AND ROCKS, ORNITHOLOGY, BIG-GAME SHOOTING, TEMPERATURES AND BAROMETRIC READINGS.

THE second letter was written to the *Leek Times*, and in the following expanded form was read at a Meeting at Swainsley of the North Staffordshire Field Naturalists' Club, when the geological and natural history specimens, as well as my Srinagar purchases in silver work, embroidery, papier-machié and wood-carving were shown and explained.

June 13, 1903.

TO THE EDITOR OF THE LEEK TIMES.

Dear Sir.—I had not enough leisure to write my second letter during my stay in Kashmir, that in which I promised to describe something of the natural history of that most interesting part of the world.

I wrote it, however, at intervals during my voyage home, principally in the Red Sea, but I leave it to your discretion to insert it or not as you may think best, as it is unavoidably rather long.

The geology of Kashmir is a big subject, and I

will only deal with a small part of it, and that principally what I saw, over but a limited area of mountain and valley, and during brief opportunities of observation combined with sport in the Western Himalayan ranges at pretty high altitudes, commencing not far from the immense springs at Achábál, Bawan and Islamabad, which mainly form the sources of the River Jhelum, which flows through the entire valley of Kashmir, east to west, through Srinagar to Baramulla, a distance of some 75 miles, bounded all the way by snow-clad mountains, with extensive valleys and nullahs running up from the plains into the heart of the great ranges. The main Valley of Kashmir is eighty to ninety miles long and over thirty broad, with numerous lateral valleys called margs, with nullahs running up into the mountains many thousands of feet high, often ending in glaciers and avalanche accumulations of snow. The whole valley is completely encircled by snow-clad ranges, with very grand and rugged skylines, escaping glaciation by their great altitudes, too high to have been shaped by other than volcanic upheaval, and subsequent atmospheric erosion. The happy valley of Kashmir averages 6,000 feet above the sea and is only approached by Tongas, along a tortuous and dangerous road two hundred miles long from Rawal Pindi, the nearest railway station. From various parts of the valley may be seen the principal mountain heights: Skardu, 8,873 feet; Nanga Parbat, 26,629; Haramukh, 16,903; Gwash Bran, 17,800; Amarnath, 17,321; the Panjal range,

15,000; Kazi Nag, 12,125, Tutakutti, 15,524 feet, etc. (see plate 30). The geology of those parts of Kashmir which I had opportunities of examining I found to be very ancient. Much of it is very difficult to correlate with corresponding epochs on this side of the globe. Yet there are analogies which show such sequence as we are accustomed to find in Great Britain, although not in the same amplication. For example, if my classification of the fossils I found is correct, there can be no doubt of a corresponding carboniferous epoch, but with an extremely limited zone. I found no signs of coal, nor of any upper carboniferous rocks, but I saw very good developments of carboniferous limestone measures, with their characteristic fossils. Of shells of Brachiopoda I found in the Panducha limestones near Singpura, Terebratula, Chonetes, Athyris, Streptorynchus, Spirifer triangulata, Producta, in which P. scabriculus is prominent, Orthoceras, and numerous Polyzoan life-forms, Polypora and Fenestella being very interesting and well developed in the shale-beds. My visit to the famous and beautiful Liddar valley was too short to give me a thorough acquaintance with its geology, for I had only time to go up a few miles, but my friend Colonel Ward, whom I had the good fortune of joining in his camp, gave me a few pseudomorphs of fossil shells in sandstone rock which appear of older date than carboniferous. They resemble very closely those of the Caradoc grits and Wenlock shales of the Lower Silurian formation of

England, both in the matrix in which they lie and in the appearance of the shells themselves.

I have written out a list of all the fossils I found. I think a few species may be considered new to English geology. I took them up to the Jermyn Street Museum and compared them with the collection there. A few of them were evidently divergencies from the typical and well-known forms, and the authorities of the museum hesitated to bestow specific names although the genus of each is sufficiently apparent. Although this is not a large collection it is I am sure of sufficient importance to interest our North Staffordshire Field Club members, some of whom I hope will be able to define or suggest new names for the species if known. A dark limestone abounds and forms the building stone of which the ancient Hindu temples of some 1,000 years ago were built; all now, sad to say, in ruins from frequent earthquakes, and still worse from the native vandalism in using their massive stones for subsequent buildings. I visited the celebrated Temple of Martand, near Islamabad, plate 31, and many others, and have brought photographs of the most interesting ones. It is a pity and a great loss to Kashmir archæology that these temples, once so fine in form, and with such characteristic details, should have been allowed to be so despoiled. Of unfossiliferous rocks I gathered specimens of schists and sandstones in various localities, notably at Srinagar; also granites, one a porphyritic greenstone, slaty and other metamorphic rocks of appar-

PLATE 31.

The Hindoo Temple of Martand, near Islạmabad, Kashmir Valley.

ent Silurian age; but the slates were without well-developed cleavage planes. There occur here and there prominent cliffs of many-bedded whitish rock, possibly Triassic, or even Tertiary, conformable with the overlying limestone and Silurian rocks. In every case they were very much folded, and gave the mountain side a very picturesque appearance. I cannot at present suggest an English equivalent, and there possibly may be none. I could not detect any break of strata between this rock and the lower Silurian measures. I had no opportunity of deciding where to correlate very thick beds of red sandstone, they may be Permian, Triassic, or Devonian, so closely do these formations lie together, and seem to pass into each other. One is led to believe that the whole series of formations from the Trias to the Silurian are here represented, but without that stratigraphical sequence and the breaks characteristic of English geology. There are red sandstones both compact and friable intercalated with the oldest rocks; in some places looking like Trias and in other places like Devonian; I could not find any fossils to identify them by. There is a singular dearth of fossils in much of this strata in this part of Kashmir. Should the shells of the Liddar Valley I have mentioned turn out not to be Silurian, then I did not observe in any of the Silurian rocks, which are of great thickness, the least trace of a fossil. Much of the Silurian measures have undergone great changes by metamorphic action, which very probably has obliterated all traces of organic remains, at least in the limited area of my examinatiou, for I searched assiduously for fossils in strata apparently thousands of feet in thickness, but found none, unless those I have just mentioned be they. These rocks seem to lie conformably under the carboniferous measures. Running into Srinagar is the end of a mountain range, and upon it, about a thousand feet high, stands the Hindoo Temple, Takht-i-Suliman, plate 16, a celebrated place of pilgrimage, said to be at least a thousand years old. It forms the most picturesque feature of Srinagar. In the centre of the temple is the customary Phallic lingam of stone with a serpent coiled round it. This part of the range consists of an enormous upheaval of amygdaloidal trap, with amygdules of chalcedon, large and small, of which, associated with and forming part of it, is a vast mass of trap rock resembling our Porphyritic greenstone, also slaty measures and large beds of unfossiliferous, argillaceous, and arenaceous softer sandstones, the exact horizon of which I am not able to determine. This immense out-pouring of igneous amygdaloid and greenstone rocks seems to have obliterated the earlier stratified measures. The carboniferous rocks, if I rightly assume them to be such, seem of scant proportions compared with those of the older series, notably those which seemed to me to correlate with our Silurian and Cambrian systems. If there has been no formation or denudation of the coal measures proper, the rest of the carboniferous rocks seem to

PLATE 16.

The Temple Takht-i-Suliman, a place of Hindoo pilgrimage.

PLATE 32.

Commencement of the Achábál Rukh or Big Game Preserve of H.H. the Maharaja.

imply that a much less time was occupied in representing this formation compared to that of Europe. There seems to be no exact equivalent and sequence to European geology. Even the apparent thickness of the stratified rocks, older than the carboniferous, appears much greater than it really is, for many of the mountain sides run parallel with the strike of the rocks, with complex foldings and complete inversions, and so much have the measures been disturbed, that it is very difficult sometimes to determine where strata do not lie bottom upward. I made for the reading of this paper a rough diagram demonstrating these inversions which although not quite geologically accurate, will indicate what I saw and my inference formed both above and below the mountain section for about three miles in length in one part of my walk along the mountain escarpments. The diagram represents the rocks dipping from south to north from a height of say 8,000 to 15,000 feet above the sea. The granite intrusion shows very plainly the main agent in the mighty effects of the disturbance, contortion and involution of the strata; dotted lines represent the inversions of the strata. A few days of rock observations extended through Achábál Rukh, (plate 32) of about five miles along the mountain side, by Islamabad and Khanabhal, the lower portions of the famed Liddar and Sind Valleys, up a large valley above Avantipura, the ancient capital of Kashmir, the Khrew Rukh,* a very mountainous part, past one

* Rukh is Kashmiri for the Maharaja's big game preserves.

of the Maharaja's best big game shooting mountains, the Valley of Arapal and Narastan to Singpura and Pampur, all on the right side of the Jhelum, and in all about thirty-five miles down to Srinagar. The only coal I saw was in Jammu, but it is a Tertiary coal, beginning to be worked in the Himalayan tract north of Jammu. It is not of good quality, but is used at the Jammu Water-works along with Tertiary Bengal coal. It is the intention of the Maharaja of Jammu and Kashmir to develop it, and steps are being taken for that purpose by a projected and much needed railway from Jammu to Srinagar, which is now being surveyed. Glaciation is an important feature at lower levels in and out of the valley. Immense deposits of huge boulders with here and there rochés moutonnés, striation, moraines and rock groovings. Since looking up what has been done by a few able geologists in the Himalayas, I find some very interesting papers by Lieut.-General MacMahon, Mr. Hudleston, Mr. Lyddeker, and especially the work of Mr. Medlicot, Captain Godwin Austen, and others, recorded in the Journals of the Geological Society and Geological Association. Notwithstanding all this recorded work, most of it of great interest, I am still of opinion that a systematic and thorough geological survey of those parts of the Himalayas in which the Valley of Kashmir is placed, is not only necessary but would prove in several ways of great usefulness. In this I am confirmed by the following quotation from MacMahon's able paper on the Geological History of the Himalayas, referring to the conflicting speculations of their age, one extremist having ventured on the pronouncement that they were first upheaved in the Eocene or even in post Tertiary periods. General MacMahon says:— "Much work in the field, combined with the skilled study of thin sections under the microscope is still needed to elucidate the age and history of the crystalline rocks of the Himalayas." Mr. Griesbach believes that the dark blue limestones of which I have brought specimens are of Devonian age, but I am firmly inclined to the opinion from the fossils they contain that they are of Carboniferous age, although I must admit that several of them appear to be of different species to any yet recorded, especially one of a large species of Polyzoa. Mr. Oldham states that throughout the whole of the Palaeozoic and Mesozoic periods the mountains of Simla and of course those of the Kashmir regions were alternately land and sea, and MacMahon thinks that during the Eocene times the sea flowed over this entire region, except over the highest snow-peaks, and it was not until the close of the Eocene epoch that the crumpling up of the strata on both sides of the ancient axis of crystalline rocks took place, and that the steady rise of the *whole* Himalayan area began and which has been going on ever since. When this period of continued elevation set in, the fiords of the Eocene sea began to shrink back from east to west; and the sea gradually retreated from the Himalayan area, and from the Punjab. He thinks that the contortion, compression and upheaval which marked the earth movements that set in at the close of the Eocene period, were connected with the intrusion of gneissose granite, but that the granite upflow did not, except in one or two instances, protrude through the surface. This granite intrusion which took place as I have said at the close of the Eocene period, greatly contorted the rocks through which it thrust itself, and greatly metamorphosed them, as may be seen in several specimens I brought home, containing fragments of these contorted rocks in the intruded granite. This stupendous action which raised the Himalayas still higher and into their present position was not a sudden outburst of volcanic force or convulsions; but is the result of a very long sustained upward pressure and a long-continued and gradual cooling of the granite. The whole encircling sky-line averaging 10,000 to 20,000 feet above the sea, is clad in the whitest of snow down to a line of about 8,000 to 9,000 feet.

The Valley of Kashmir is a flat plain of pleistocene alluvium, and is composed of loams, clays and sands. It is the ancient bottom of a lake, which formerly occupied the whole area of the valley to a depth of perhaps more than a thousand feet, and is said to have found its outlet by volcanic agency through the narrow gorge at Baramulla. It is interspersed with long flat-topped banks and ridges of detrital wash, called karewas, which consists of more or less fertile and unfertile alluvium washed down by mountain and lake currents, probably during the existence of the great lake itself. From these deposits the sun-dried brick walls of the houses are built, and coarse porous pottery is made. The whole country is a mass of volcanic disturbance, ancient and modern; much of it contemporaneous with, as well as prior to, the carboniferous epoch, as well as subsequently. Kashmir is situated in a line of seismic weakness. Earthquakes are still frequent; a severe shock occurred in April during my stay. In 1873 an earthquake destroyed 3,000 people and large numbers of cattle, and no year passes without recurrences of greater or less severity.

The rocks surrounding this great lacustrine basin are mostly stratified, except where absolutely igneous.

This collection of Kashmir rock and palæontological specimens, although small, may prove of interest to our North Staffordshire Field Club on some future occasion. I am happy to state that it has been accepted by Professor T. McKenny Hughes, the eminent Geologist and Principal of the new Sedgwick Museum, Cambridge and will find a home in that important geological collection.

Illustrations of some of the principal fossils described will be found in plate 41, the drawings of which have been beautifully executed by my niece Miss Elinor Wardle, and forms one of the most important and interesting of the illustrations.

List of Specimens of some Kashmir Rocks and Fossils I collected in Kashmir, in May, 1903.

No. 1.—Large block of Limestone with Polyzoa, Producta, Rhynconella spondi, Polypora Rhabdomeson, Glauconoma, etc., from Panduchuck, near Singpura, upper side.

,, 2.—Ditto lower side.

,, 3.—Limestone from Panduchuck, with Avicula, Polypora, Polyzoa, etc.

,, 4.—Ditto Polyzoa (plate 41, fig. 1).

,, 4A.—Ditto ditto magnified (plate 41, fig. 1a).

,, 5.—Ditto Polyzoon, resembling Polypora (plate 41, fig. 2).

,, 6.—Ditto Crinoids.

,, 7.—Ditto Retzia.

,, 8.—Ditto Polypora, Crinoidea, Rhabdomeson.

,, 9.—Ditto Brachiopoda, probably Rhynconella (plate 41, fig. 7).

,, 10.—Ditto Brachiopodas, probably Rhynchonella (plate 41, fig. 8).

,, 11.—Ditto Brachiopodas, Strophomena, Spirifera.

,, 12.—Ditto Spirifera (plate 41, figs. 4, 6 and 9).

,, 13.—Ditto Polypora, Diastopora, etc.

,, 14.—Ditto Producta, Brachiopoda, etc.

,, 14a.—Ditto Productus, allied to pustulous.

,, 15.—Ditto Productus, Polyzoon (plate 41, fig. 3).

PLATE 41.

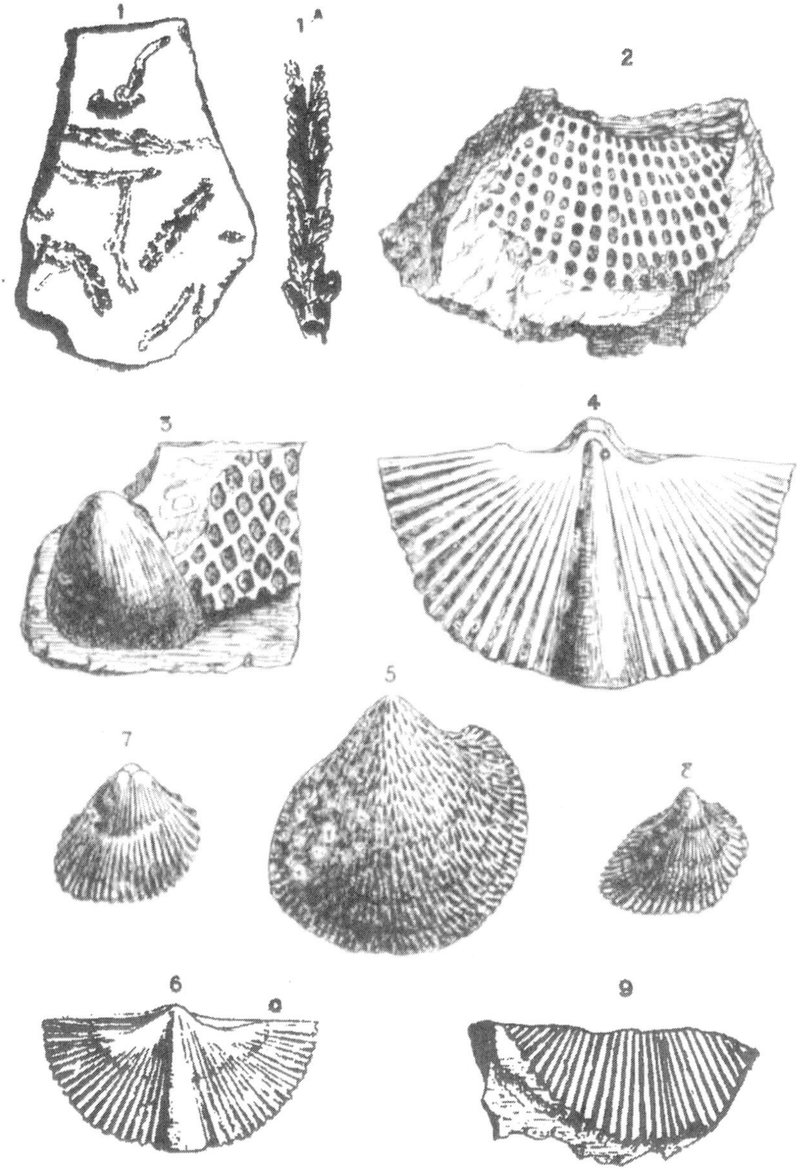

A few of the Fossils collected at Singpura, Avantipura, Liddar Valley, Srinagar, etc.

No. 16.—Ditto Streptorynchus and Polyzoa.

,, 17.—Ditto Productus, allied to P. pustulous (plate 41, fig. 5).

,, 18.—Millstone or probably Silurian Grit, resembling Caradoc grit, Liddar Valley, Pseudomorphs of Spirifera, a Conchifera and Polyzoa.

,, 19.—Ditto, ditto, Brachiopoda, Rhynchonella.

,, 20.—Ditto Pseudomorphs, fragment of Brachiopoda.

,, 21.—Ditto fragment of Brachiopoda and Monticulipora.

,, 22.—Ditto Spirifera, etc., Polypora.

,, 23 to 27.—Series of Limestones in succession, near Avantipura, the white looking ones very much contorted, squeezed and compacted with very fine greenish quartz felsite.

,, 28.—Porphoritic Greenstone, from Singpura, close to the carboniferous limestone fossils and limestone series of Nos. 23 to 27.

,, 29.—Carboniferous Limestone, of which the ruined Hindu Temple at Martand, near Islamabad, was built.

,, 30.—Igneous Rock, near Avantipura, of typical quartz-felspar porphyry related to No. 28 and belongs to the same series of intrusions, connected with some granite below or in the area.

,, 31.—Ditto, but with less quartz. These are probably dykes or tills.

No. 32.—Amygdaloid igneous Rock from base of the mountain at Srinagar, possibly a lava and contemporaneous with the sedimentary measures.

,, 33.—Slaty Rock from base of mountain at Srinagar.

,, 34.—Limestone from a stratified cliff at Bawan, Liddar Valley.

,, 35.—Ditto, ditto, ditto.

,, 36.—Green and purple Rocks from Uri, probably calcareous shales; one of these would make a good section to show the effects of the shearing stresses on these rocks.

,, 37.—Road-mending Carboniferous Limestone near Murree, 7,000 feet above sea level.

,, 38.—Amygdaloidal igneous rock from Arapal, probably the fine edge of a rock like No. 32.

During my fifty-one days stay in Kashmir, I had the honour of being entertained as the State guest of the Maharaja of Jammu and Kashmir, of whose kindness and hospitality it is not possible to speak too highly. During the time I was in Jammu he placed one of his splendid elephants and a carriage and pair at my service. The elephant was one that attracted attention at the Delhi Durbar, one of the finest specimens of its kind, with head and front richly and decoratively painted (plate 9).

His Highness the Maharaja and his brother, General Raja Sir Amar Singh, were most courteous,

PLATE 40.

Author on pony, and Camping Staff.

thanking me most warmly in Durbar assembled at the Palace for the interest I had taken in establishing Kashmir Sericiculture, and for my continued attention to it. They invited me to shoot in their special reserves, or Rukhs, as they are termed.

But my time was chiefly occupied with matters connected with the production and reeling of raw-silk, with a view of its improvement, and in reporting on its present state and future developments. I was, however, able to devote nearly a fortnight to natural history work and shooting.

On my arrival at Srinagar from Jammu, I found arranged for me, by instructions to the Governor of Kashmir from His Highness the Maharaja, a house-boat on the Jhelum, a covered Shikara, and two Doongas, one for carrying tent appliances, and the other for cooking. These were for my expedition upon and up the river.

To garrison this expedition I was provided with a pony and Groom (Syce) see plate 40, on the right the Cook (Khansama); my native servant, (Bearer or Boy); Waiter (Karthmagar); Boat Captain (Jamadar); Tent man (Frash); and in additon twelve boatmen.

At the close of the first day's journey up the Jhelum the boats were roped to the river side, and the boatmen applied for money to buy food in the adjoining village. On my being told by my native servant that two annas (2d.) per man would suffice, I replied that I supposed it was for the evening meal; he said, "No, it is enough for the whole day, and they will be well satisfied," and during my week's stay on the shooting tour it only cost me two shillings per day for the twelve boatmen. It is customary for the State guest to provide the commissariat of such expeditions.

Srinagar and the Jhelum constitute a veritable second Venice, with its house-boat life and charming river scenery.

I had the good fortune to shoot two fine stags, Bárásinghás, (Cervus Kashmeriensis) a wild boar, (Sus Cristatus) in the Achábál Rukh, and a black bear (Bomba Hapat) in the Khrewe Rukh (plate 42). This was skinned in the evening, and by seven o'clock the next morning I counted thirteen immense vultures devouring the carcase.

Three days spent in the Achábál reserve, driving and stalking for Bárásinghás and bears, gave ample opportunities for geological observation from the base of the mountain to a height of about 9,000 to 10,000 feet, parts of which ascents were on pony-back nearly half-way up, and after that on foot, the ascents being too steep for pony climbing, so steep in some parts that it was difficult to hold on except by the rock stratification ledges. I got to consider an angle of forty-five degrees a gentle slope.

A coolie carried my hammer and specimen bag, another my shot-gun, my shikari carried my rifle, and I supported myself with a stiff alpenstock, or rather Himalayan-stock.

Plate II.

PLATE 42.

Souvenirs of Sport shot in the Himalayas by the Author, during his visit to Kashmir in 1903, also Curios collected.

Mahomedan Mosque, Shahihumdan at Srinagar, on the Jhelum.

On reaching one of the Himalayan ridges I tracked in the snow for about 200 yards the fresh footsteps of a leopard, but failed to find him, although the nullahs were well beaten by a staff of coolies. He had gone farther afield.

I shot a collection of birds, chiefly under the guidance of Colonel Ward, the authority on the Fauna of Kashmir, who kindly invited me to accompany him. He has the oversight of the big game and the Rukhs of Kashmir, and I shall not forget the pleasant week I passed with him and Mrs. Ward, both on the Jhelum and in camp. I am indebted to him for the names of the following birds I shot, whose skins, with others, I am bringing home:—

Black stork (Ciconia Nigra), coot, large (Fuleca atra), western spotted forktail (Henicurus Maculatus), common kestrel (Cerchneis Tinnunculus), koklas pheasant (Pucrasin Macrotopta), dabchick (Podiceps Phillipensis), yellow-headed wagtail (Motacilla citeoloides), Himalayan goldfinch (Carduelis caniceps), common sparrow (Passer communis), cinnamon sparrow (Passer cinnamomeus), the rufous-backed shrike (Lanius erythnonotus), the black and yellow crossbeak (Hesperiphona icterioides), white-cheeked bulbul (Otocompsa leucogenys), the Himalayan babbling thrush (Trochalopteron Simile), the Himalayan whistling thrush (Merula unicolar), the Himalayan pied woodpecker (Picus Himalayanus), the Western Himalayan pied woodpecker (Dendrocopus Himalayensis), hoopoe (Upupa Epops), wren

(Anorthura neglecta), woodcock (Scolopax rusticula), the Chikor partridge (Caccabis chukar), large spotted nutcracker (Nucipago Multipunctata), king crow (Decrurusater, or Black Drongo), Monaul pheasant (Lophophorus refulgens), the Himalayan tree creeper (Certhia Himalayana), the short-billed minevet (Pericrocotus brevirostris), the white-capped redstart (Chimorrornis leucocephala), the Indian Paradise flycatcher (Terpsiphone paradisi), the mango bird (Oriolus Kundoo), Imperial eagle (Aquila Heliaca), Indian bush-chat (Pratincola Maura), nut-cracker, tit-mice (several species), and tits.

The sparrows seem to be the most numerous species of bird; next come jackdaws, rooks, and crows; after them the Mynah. Swallows and martins were abundant.

Of the trees of Kashmir the following predominate: of fruit trees, glorious in blossom, peach, almond, apple, pear, plum, apricot, and cherry. The mulberry luxuriates all over the valley and is indigenous, in some places it occurs in forest quantities.

The willow, well pollarded, abounds. The finest trees are the chenars or planes in every village, some of them two hundred feet high. I measured he base of one of the largest, it was twenty yards in circumference. The leaves of the chenar constitute the chief motives in their beautiful designs for silverwork, wood-carving, embroidery, etc. Deodars, spruce, and other conifers abound on the mountain slopes amongst the snow.

The mulberry is very badly forested, and in my

PLATE 13.

Dhunga and Houseboat. Colonel Ward and Author.

Report I advised the appointment of a forestry expert; this will be done from the Indian Forest Department. The trees have been cruelly lopped and denuded of their boughs in many places, and years will elapse before many of them will be productive of leaves for silkworm food, but both the black and white mulberry are being extensively planted. In the Peshawar and Srinagar dyehouses I found a few interesting dyes. Fast yellow is dyed with barras, or barruc, imported from Cabul; also a yellow dye called mazath, or mazate, the root of the wild geranium; another yellow dye called rawan chini; reds, from cochineal; blues, indigo.

Of Kashmir flowers, they have only to be seen. The entire valley was teeming with fruit blossoms, and the sight of the wild iris, and the wild crown Imperial lily, as well as many other flowers, was of indescribable beauty and interest.

It is impossible to omit mentioning in this letter the exceedingly pleasant recollections I have of the delightful hospitality I received from Mr. and Mrs. Colvin, at each Residency, at Sialkot, Jammu, and Srinagar, and also of the kindness shown to me by Mr. and Mrs. Walton during my stay in Srinagar, a few days of which I had the pleasure of enjoying their hospitality.

It is fortunate for Sericiculture in Kashmir that Mr. Colvin, the Resident, and Representative of the Government of India, is taking such a keen interest in its progress.

It fell to my lot, at the request of the Government of India, to recommend a Director of this commencing Industry in 1897. The splendid results shown since that time prove that the appointment of Mr. Walton, who for many years previously had much experience in the Bengal silk districts, has been justified, and that under his able guidance and control, the industry has yet most promising prospects, and I am glad that I am able to report to the Government of India my entire satisfaction at the present state and future prospects of this important industry.

Knowing I should have to experience much change of temperature in the various parts of my journey to Kashmir, I carefully took both temperature and barometric readings daily, and often two and three times daily. In all I took 179 readings, the most interesting of which I give in the following table. Of course the great differences in the barometric readings were caused by changes of altitude rather than by the weather changes, for example, at Murree, where I had to pass through the snow on foot at a height of 7200 feet, one third of the way between Rawal Pindi in the Punjab plain of India, and Baramulla, the entrance into the Valley of Kashmir.

TEMPERATURES AND BAROMETRIC PRESSURES

FROM 13TH FEBRUARY TO 28TH MAY, 1903.

Place.		*Date.*	*Hour.*	*Temperature.*	*Barometer.*
		1903.			
Leek ...	...	13 Feb.	9 a.m.	45	29·7
London	...	13 ,,	6 p.m.	48	30·62
Paris ...	...	14 ,,	6 p.m.	50	30·0
Alais ...	...	15 ,,	8 a.m.	52	29·5
Marseilles	...	20 ,,	9 a.m.	55	30·0
Port Said	...	25 ,,	noon	70	30·45
Suez ...	...	26 ,,	4 p.m.	78	30·3
Red Sea	...	27 ,,	8 a.m.	73	30·3
Do.	...	27 ,,	4 p.m. (deck)	75	30·3
Red Sea	...	27 Feb.	7 p.m. (cabin)	82	30·20
Do.	...	28 ,,	6 p.m. (berth)	80	30·20
Indian Ocean		1 Mar.	1-30 p.m.	78	30·15
Do.	...	5 ,,	8 a.m.	76	30·15
Bombay	...	7 ,,	11 p.m.	76	30·10
Do.	...	9 ,,	5-30 p.m.	90	30-10
Bhopal	...	10 ,,	8 a.m.	80	
Gwalior	..	10 ,,	4-30 p m.	86	
Delhi	...	11 ,,	8 a.m.	64	
Umritza	..	11 ,,	11-30 a.m.	62	29·25
Sialkot	...	12 ,,		64	29·6
Jammu	...	13 ,,	9-49 p.m.	57	29
Do.	...	15 ,,	8 a.m.	60	28·95
Do.	...	15 ,,	3 p.m.	68	28·9
Peshawar	...	18 ,,	8 a.m.	62	29
Do.	..	20 ,,	8 a.m.	64	28·76
Ali Musjid	...	20 ,,	9-18 p.m.	66	28·77

Place.	Date.	Hour.	Temperature.	Barometer.
	1903.			
Rawal Pindi	21 March	6 a.m.	60	28·35
Trez, en route to Kashmir	21 ,,	7-3 p.m.	52	26·75
Dulai... ...	22 ,,	7 a.m.	52	26·6
Murree ...	22 ,,	12 noon	42	24·8
Do. ...	22 ,,	4 p.m.	36	25·5
Dulai	22 ,,	8-30 p.m.	56	27·60
Guori ...	23 ,,	10-45 a.m.	54	27·45
Uri	23 ,,	4-30 p.m.	50	25·7
Baramulla ...	23 ,,	9 p.m.	38	25·15
Srinagar ...	Average from 25 March to April 1	8 a.m.	45	25·0
Acáhábl ...	Average of 4 days at	8 p.m.	44	24·9
Khanabal ...	6 April	8 a.m.	42	25·1
Srinagar ...	6 ,, to 17	7 p.m.	51	25·1
Singpura ...	17 ,,	7 a.m.	55	25
Avantipura ...	21 ,,	7 a.m.	51	24·81
Srinagar ...	26 ,,	7 a.m.	52	24·93
Baramulla ...	2 May	noon	60	24·96
Uri	2 ,,	4-25 p.m.	58	25·4
Chakoti Dak Bungalow ..	2 ,,	6-30 p.m.	58	25·95
Garhi ,, ...	2 ,,	9-30 p.m.	60	27·05
Do. ...	3 ,,	6 a.m.	64	27·15
Domel ,, ...	3 ,,	8-20 a.m.		27·52
Dulai ,, ...	3 ,,	9-30 a.m.	72	27·77
Rawal Pindi	4 ,,	8 a.m.	79	28·2
Lahore ...	6 ,,	9-40 a.m.	90	

Place.	Date.	Hour.	Temperature.	Barometer.
	1903.			
Agra, in train	6 May	8 a.m.	85	29·35
Gwalior ,,	6 ,,	12 noon	104	28·8
Do. ,,	6 ,,	4 p.m.	106	28·8
Munmar ,,	7 ,,	8 a.m.	90	28·1
Summit Station, in train ...	7 ,,	8-30 a.m.	90	27·95
Kalyan, in train	7 ,,	1 p.m.	102	29·9
Bombay ...	8 ,,	8 a.m.	88	29·05
At Sea				
Bombay ...	9 ,,	5-30 p.m.	86	29·9
Indian Ocean	10 ,,	10 a.m.	84	30·025
Do. ...	11 ,,	8 a.m. (cabin)	86	30·02
Arabian Sea...	12 ,,	7 a.m.	84	30·08
Nearing Aden	12 ,,	9-30 p.m.	92	30·10
Aden	14 ,,	7 a.m.	86	30·1
Do. ..	14 ,,	2 p.m.	90	30·1
Do. ...	14 ,,	10 p.m.	88	30·1
Red Sea ..	15 ,,	7 a.m. (cabin)	87	29·05
Gulf of Aden	17 ,,	7 p.m.	80	30
Suez Canal ...	18 ,,	7 a.m.	72	30·25
Port Said ..	18 ,,	10 p.m.	78	30·22
Mediterranean	19 ,,	7 a.m.	73	30·22
Do. nearing Italy ...	21 ,,	8 a.m.	69	30·35
Straits of Messina ...	21 ,,	7 p.m.	71	30·36
Nearing Corsica	22 ,,	7 a.m.	67	30·45
Do. ...	22 ,,	8 p.m.	68	30·50
Marseilles ...	23 ,,	6-30 a.m.	70	30·5

Place.		*Date.*	*Hour.*	*Temperature.*	*Barometer.*
		1903.			
Lyons	...	25 May	8 a.m.	74	29·8
Paris . .	...	26 ,,	8 a.m.	72	29·6
Boulogne	...	26 ,,	noon	75	29·8
Calais	...	26 ,,	1-5 p.m.	66	30·0
London	...	26 ,,	12 noon	64	30·45
Leek ...	...	28 ,,	8 a.m.	60	33·25

The stopping places from Rawal Pindi en route to Srinagar are eleven in number, at each of which is a rest-house or Dak Bungalow for travellers and horses, at very cheap rates. The journey generally takes four days, and the distance is nearly two hundred miles. I did the whole journey back in Tonga in two long days, changing horses about every six or seven miles, nearly all downhill.

The Dak Bungalows are at the following places, with their distances apart from each other :—Rawal Pindi to Barakao, 13½ miles, road level; Tret, 12 miles, 4,000 feet above sea; Murree, 13½ miles, 7,200 feet above sea; Kohala, 27 miles, 2,050 feet ab ve sea; Dulai, 12 miles; Domel, 9½ miles; Garhi, 13 miles; Halli, 10 miles, 3,000 feet above sea; Chukoti, 11 miles; Uri, 13½ miles, 4.000 feet above sea; Rampur, 13 miles; Baramulla, 14 miles; and Srinagar, 30 miles.

The climate of Kashmir in the summer and autumn is delightful, and very nearly that of Italy. The mean temperature in July is about 75 degrees. The annual rainfall is about 25 inches, and is much less than other Himalayan stations. The winter is

PLATE 14.

English Church at Srinagar; Vicar, the Rev. C. E. Tyndale-Biscoe.

Assistant Masters and Pupil Teachers of the Rev. C. E. Tyndale-Biscoe's Church Missionary Society School in Srinagar.

severe, ending in March. Snow falls in December with keen frost. In January more snow falls, which remains for about three months, during which time the valley is covered to a depth of nearly a foot.

Last but not least I must not omit to record my pleasant impressions of the admirable work that is being done in Srinagar by the ladies of the Church of England Zenana Mission, of whom Miss Hull, a sister of one of our ablest English Geologists, my friend Dr. Hull, F.R.S., is the head. Amongst other publications, she is the author of the charming little book entitled, "Vignettes of Kashmir."

Also the splendid work of the Church Missionary Society, under the guidance of the Rev. C. E. Tyndale-Biscoe and his coadjutors. He is also the Vicar of the English Church at Srinagar (see plate 14.)

His Schools are wonderfully popular, both amongst the Mahomedans and Hindoos. He has upwards of one thousand pupils, and I have much pleasure in reproducing two photographs, one showing the Assistant-Masters and Pupil Teachers of the Educational Department (plate 15), and another many of the Pupils and Teachers; in this the Rev. C. E. Tyndale-Biscoe is sitting, cap in hand, on the middle verandah roof on the right; Mrs. Biscoe and other ladies below (plate 17). Both these plates are reproductions of excellent photographs taken by my friend Geoffrey W. Millais, Esq., famous for his Kashmir photography, whom I had the pleasure of meeting in Srinagar at the time, and who has kindly given me permission to copy them.

Yours faithfully,

THOMAS WARDLE.

Leek, 8th June, 1903.

PLATE 17.

Rev. C. E. Tyndale-Biscoe and one of his Schools.

CHAPTER XXI.

ON BENGAL SERICICULTURE.

COPY OF THE PROCEEDINGS OF A CONFERENCE ON SILK HELD IN THE OFFICE OF THE REVENUE AND AGRICULTURAL DEPARTMENT OF THE GOVERNMENT OF INDIA AT CALCUTTA, ON THE 8TH JANUARY, 1886.

A MEETING was held at the office of the Revenue and Agricultural Department on the above date for the purpose of discussing questions connected with silk production, and with the special measures to be taken for the equipment of a special Silk Court at the forthcoming Indian and Colonial Exhibition. Mr. E. C. (now Sir Edward) Buck, Secretary to the Government of India, Revenue and Agricultural Department, presided, and Mr. Wardle, who had just returned to Calcutta after visiting the Eastern Bengal silk districts, took the leading part in the proceedings.

Mr. Buck introduced Mr. Wardle to the Conference as a gentleman who had come out from England to help the Royal Commission to prepare for the Indo-Colonial Exhibition such a collection of silks and information as might tend to awaken the attention of manufacturers. When the Exhibition was first announced the Government of India was reviewing the various products of the country with the idea of determining which might be the most likely to reward any effort which might be made to promote its development, and decided that silk was of all the products of India the one of which there was most hope. Holding that opinion, it had allotted a larger portion of the Exhibition space at its disposal to a complete collection of silks than to any other class. During the year, however, information was received from the Royal Commissioners that they were extremely anxious the Exhibition should be utilised for the purpose of bringing prominently forward the question of silk production to the notice of the public, and it was, by the mutual wish of the Royal Commission and of the Government of India, that Mr. Wardle, at the sacrifice of much time and personal comfort, consented to come out to India to do what he could in the same cause. It was at the same time suggested that there should be a special Silk Court, of which Mr. Wardle would have charge, and the proposal was put before the Bengal Silk Committee; for, before the question of Mr. Wardle's coming out here had been raised, the Bengal Government had been asked whether they would undertake the preparation of the Silk section, seeing that, of all the provinces of India, Bengal was the most interested in the silk question. The Bengal Government agreed to a Committee being formed in Calcutta, and the matter had proceeded so far when Mr. Wardle's arrival was announced. The Bengal Government readily agreed to the proposal for a Silk Court and offered their collections and assistance to Mr. Wardle.

Mr. Wardle then addressed the meeting. He began by saying that it might naturally be asked why he, as a silk dyer and silk and calico printer, should appear there to speak upon silk cultivation and the earlier operations in the manipulations of silk.

He would anticipate objections of that kind by showing that, apart from lengthened scientific investigation, he might be considered to have at least some claim to be heard on the subject from having been accustomed as a silk dyer and printer during his business life, to examine every variety of silk fibre and its diversified kinds of manufacture, and had therefore had lengthened opportunities of studying the nature and many forms of the silk which came to him to be dyed or printed.

He hoped he could at least from this long acquaintance speak thoughtfully on the subject, and he must add that he spoke also from a disinterested point of view.

Having as early as 1872 succeeded in making some important developments in bleaching and dyeing Tussur silk, at that time a very difficult fibre to dye, he had been consulted shortly afterwards by the Marquis of Salisbury, then Secretary of State for India, to report on the practicability of dyeing this silk in India and in examining the native dyes of India. The investigation took eight years to complete, during which time he was requested by the Government of India to take charge of Indian silks at the Paris Exhibition of 1878, and now he came out to India at the request of the Royal Commission of the Colonial and Indian Exhibition of this year to undertake the same duty, but on a more extended scale, and to properly emphasise the capabilities of India as a silk-producing country. Naturally with so long an acquaintanceship with both the domesticated and the wild silks of India he had their progress and utilities very much at heart.

Further, he was led to accept the invitation at Mr. Buck's expressed wish that he should come out in the interests of Indian sericiculture.

He was here for several specific reasons; *first*, to stimulate and give encouragement to an extended production of Tussur silk, now a settled European manufacture and industry, requiring a much larger supply of the raw material than India was at present able to supply; *secondly*, to collect examples of the wild and domesticated silks of India in all their stages, from the eggs of the various silkworms to the finished fabrics for the Exhibition; *thirdly*, to see for himself how it was that Bengal silks had gradually fallen into disrepute in England and in all the other European silk centres during the last twenty-five years until they now hardly held a place equal to their former reputation.

Also to make an examination on the spot of the present system of reeling silk and to point out how it might be improved, and if possible, made equal to that of the best silk-producing countries.

Previous to that time in the town of Leek in which he lived, the staple trade of which was and is the manufacture of sewing-silks, the consumption of Bengal silk was the rule, that of other silks being the exception. Bengal silk, if properly reeled, has several properties which peculiarly fit it for sewing-silk, in addition to its suitability, when properly reeled, for weaving purposes, and a large quantity could be used for that important requirement.

Gradually the silks of China and Japan came in, and for many years past had almost entirely swept away the use of Bengal silk, and his motive was to try and restore this silk to its former position, one which he should be able to show from his recent visits to the silk filatures of Bengal, might easily be more than realized. He remembered that two great defects existed in Bengal silk, which, gradually increasing, had gradually led to its being in less demand, for many of the important purposes for which it had been used. One was the very defective way it was reeled from the cocoon, and the other the difficulty there was in winding it when reeled. He was disposed to think, after seeing the rough-and-ready way the natives reeled their cocoons in their village-huts, that native-reeled silk had too frequently been shipped to Europe instead of that of the better reeled Indo-European filatures, but no doubt even the high quality of the latter, although beyond all comparison with native reeling and also from other reasons, had fallen from its former excellence.

These were the principal causes which led to the decadence of the Indian trade not only in England but also in the other European silk manufacturing centres. From lengthened microscopical and other examinations of Bengal silk he had convinced himself that the fault was not in the fibre itself, for in comparison even with Italian silk its structure left nothing to be desired.

The principal cause of the irregularities in reeled silk is a defect termed by the French and Italians *duvet*, the precise nature of which he was told in all the filatures he had visited both in Italy and India he had been the first to point out and define. It was strange that this defect had not before been thoroughly understood, but as the use of the microscope in Indian sericiculture had hardly yet commenced, no one need be surprised at erroneous conclusions being drawn in matters almost invisible to the unassisted eye. He could not too strongly advocate the constant use of the microscope in every filature. It was not necessary to have an expensive instrument; he had brought out with him one of Beck's new small £4 4s. 0d. microscopes with objectives of 1 inch and ¼ of an inch power, quite sufficient for all practical purposes. To this should be added a micrometer eye-piece for making measurements of the thickness or diameter of the fibre.

Many opinions of the supposed nature of the *duvet* were given him by the several gentlemen

PLATE 18.

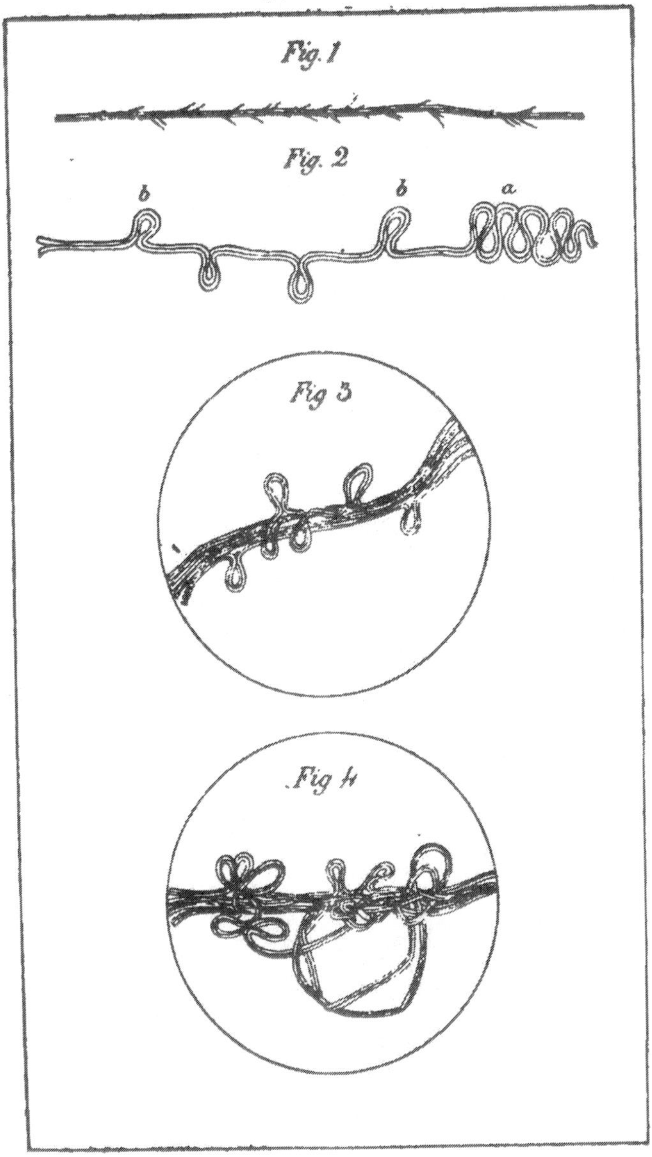

Cocoon Fibre, showing Bave and Duvet.

engaged in the filatures both in Italy and India. Amongst them were these principal ones, *first*, that the *duvet* consists of broken ends of the *bave* arising from the piecing of the thread by the reeler or by replacing carelessly the end of a fresh cocoon; *secondly*, that the silkworm does not deposit its silk in a continuous thread but leaves off and recommences at intervals throughout the cocoon, causing ends in its double parallel thread or *bave*; *thirdly*, that the fibre is hairy, somewhat as in fig. 1, plate 18; but this is not so. The silkworm deposits its silk with extraordinary regularity and method. The outer and inner threads of the cocoon are, indeed, finer than the central portions, the thickness increasing very gradually to about the middle of the cocoon, but the thread is, in every case he had examined, always continuous and of excellent regularity. He had never been able to detect the *bave* or fibre of any silk to be hairy or externally fibrous. It is in the *reeling* only that defects begin to appear.

The silkworm, by moving its head from one side to the other, deposits its double thread in loops as at *a* in fig. 2, plate 18.

At the points of contact the gum adheres more firmly than it does laterally, consequently in the reeling the *bave* frequently comes off in loops as at b . . . . b fig. 2, plate 18.

In fig. 3, plate 18, is given the appearance of *duvet* upon a thread of raw-silk composed of four cocoons reeled together.

Fig. 4, plate 18, is the microscopical appearance of an aggravated form of *duvet*, in which a series or layerlet of loops have come off together, and which having passed both *filière* and *croissure* without being reduced, have resolved themselves into a tangle which constitutes what is known as "knibbiness" or "bouchons" in the raw-silk.

In native reeling it is much worse and much more frequent. In the Bengal cocoon this defect is much aggravated from two causes—

1st, by the greater amount of gum or *grès* which the worm exudes to cover its silken thread than in those of the Italian, French, Chinese, and Japanese worms.

2ndly, from the Bengal cocoon being smaller than those of the countries just mentioned. This smallness increases the difficulty of reeling, especially towards the inner portion of the cocoon.

In the Italian cocoon the chrysalis is larger than the Bengal ones, and helps by its greater weight to prevent the cocoon being drawn out of the water in which it floats whilst being reeled. In the Bengal cocoon when the *bave* is nearly all reeled, it is often jerked from the water to the *filière*, many loops being loosed and drawn up on the reel at once. Often for want of care whole layers come off at once and form what are called "bouchons," "knibs," "foul," "slubs," &c. The cocoon, it should be stated, is as stratified a piece of construction in its way as the liassic formation, and the layers are very apt to come off *en masse*. This is particularly noticeable in village native reeling, where irregularity of thread, however great, seems to be no drawback. He had seen the silk drawn off in positive lumps, and the wonder was that with the extraordinary rapidity of their reeling they obtain any but a thread of *chassum*, so clumsily and rapidly do they piece up the breakages, unavoidable in their rough-and-ready reeling.

In the best European filatures the irregularities are merely the occurrence of loops, *duvet* when only a few occur. "Knibs," "slubs," "foul" or "bouchons," when in quantity or in mass, are but very seldom found in the best reeled silk of Italy.

He had suggested that attention should be directed to a better and more perfect state of tension between the bassine and *filière*, and before the thread reaches the Tavelette, for no amount of *croissure* can reduce the *duvet* of a single *bave*; *i.e.*, before the cocoon fibres join together, as it is hopeless to attempt to reduce the loop on a single *bave* when it has passed through the hole of the *filière* which unites into one thread the *baves* of as many cocoons as may be reeled together—

Of Italian silk 4 cocoons reeled together are required to make a thread of 9 to 11 deniers;

and 10 to 12 to make a thread of 25 to 30 deniers.

Of Indian tussur silk 4 cocoons are required to make a thread of 40 deniers;

and three for 30 deniers.

He was cordially thanked in Italy for having pointed out the nature of the *duvet* and in having shown that by the slightest pull of the single *bave* this *duvet* was resolved into a perfect line, and steps are being taken to overcome the defect.

The *duvet* is not easily seen on the Italian silk except when on the reel and in a good north light, but its removal will mark a new era in all *Bombyx Mori* silks, particularly that of Bengal, in which they have hitherto been more numerous.

Referring to his visit of the past ten days to the Bengal silk districts, he said that at the invitation of Mr. J. J. J. Keswick and Mr. Lyall, he had gone through the filatures in the Rajshahye and Murshidabad districts. He had been very much impressed with his visit. He found beautiful raw-silks, although capable of improvement, being produced all over the district, and he had been received with the greatest courtesy and kindness; and while there, he had drawn up a few suggestions, which would, he hoped, be of special interest to Mr. Buck, as Secretary to the Government of India, in the Revenue and Agricultural Department. He had visited Mr. Morey, of the Surdah Filature, also Mr. Stocks, of the Bengal Silk Company at Murshidabad, and subsequently Messrs. Louis Payen & Co., meeting Monsieur Gallois. With regard to the properties, the structure, and the fibre of the Bengal silk, he had shown them side by side under the microscope with Italian silk, and proved to these gentlemen the accuracy of his former examinations, that there was really no practical difference in the fibre, and that of the two, the Bengal silk might be preferred in several respects if it was properly reeled. It was surprising how near they were together both in size and homogeneity of thread. He was, however, bound to remark that the present mode of reeling in the Bengal filatures was a little old-fashioned, and he thought it was worth while to invest money in altering it, and to try the Italian mode of reeling. He was not sure whether he knew enough about the matter, but on his way to India he had visited Italy, and had seen there what he considered to be the very best method of reeling. He saw every detail in connection with the preparation of the cocoons and the subsequent reeling of the silk. He also saw the working of the reeling-machine known as the *Tavelette Consono*, a simplified and beautiful contrivance exactly suited to the reeling of the *finest* silks. Although the Italian and Indian worms were of the same genus there was a specific difference between them.* The Italian worm at full size was much larger than the Indian one. Mr. Wardle would not now enter into scientific considerations on the point, but he was quite certain, and had convinced the gentlemen he had visited, that the silk of the Indian worm was as good structurally as the Italian one. The Bengal cocoon was not as large as the Italian, and did not con-

* The European, Japanese and Chinese univoltine silkworm of commerce is *Bombyx mori*, the multivoltine ones of Bengal are *Bombyx fortunatus* and *Bombyx crœsi*.

tain so much silk. From measurements, he found that the white and yellow cocoons of the old Italian breed measured 650 metres in length each, the Bengal cocoon only averaging 150 metres. This was an immense difference. There was some cause for this, and he thought it might result from the Indian worms being multivoltine, and also from their being imperfectly bred and badly nourished. He had been at some trouble to obtain the matured opinions of the gentlemen engaged in the industry in Eastern Bengal, and he would place a summary of them before the meeting. On the authority of each of the filatures he visited, the rent exacted by the zamindars from the ryots who cultivate the mulberry plant is excessive. Before the American War it was Rs. 1-8 to Rs. 2 per bigha. During the war and the prosperous few years which succeeded, the rent became Rs. 14 and even as high as Rs. 16 per bigha. In the neighbourhood of Berhampur it is now Rs. 12 to Rs. 14; in Malda Rs. 16. There are three bighas in an English acre, and the rent would be Rs. 48 per acre—a rent very much higher than the average of English rents. Rather than lower their rents, the zamindars would sometimes allow their land to lie fallow. In the good years he had just spoken of, the ryots paid any price the zamindars asked, and it seems to be nothing short of a misfortune that rents have not come down with the depression of recent years, and the declining demand for Bengal silks.*

* Since the meeting he had received the following letter

It is the native zamindar who exacts such very high rents and not the European,† and who, if he would wish to see the industry again in a flourishing state, would do well to be wise in time, and assist the present effort of its revival by

from a native gentleman occupied with native interests but who did not wish his name to appear:—

"Dear Sir,—Government holds certain lands as intermediate proprietor or zamindar. These lands or estates are called *Khas mahals.* Government as a model zamindar can introduce in the *Khas mahals* all improvements and reforms which are likely to increase the material wealth of the country. The ryots hold land immediately under the Government and, while letting out land, the Government can make its own terms with them. For instance, if there be any mulberry lands in any one of the Government *Khas mahals*, and if the Government be convinced that low rate of rent is likely to improve the quality of cocoons grown and so foster the trade in silk, the Government can very easily set an example by reducing the rent in the *Khas mahals.* In fact, as a model zamindar, Government can introduce any reforms in the *Khas mahals* that are likely to develop agriculture, trade, or industry.

"In the Courts of Wards' Estates, Government has no such free scope as in the *Khas mahals.* In those estates, the Government holds the land temporarily in trust and can within a certain limit introduce reforms which the actual proprietor would have introduced had he been in possession. These estates are generally the estates of minors, and Government holds or manages them in trust for them.

"You would confer a great boon to the country if you could prevail on the Government to act as a model zamindar in these *Khas mahals* and show the poor ryots what can be done in the way of improving agriculture, trade, or industry."

Yours faithfully,"

---

† My correspondent in conversation rather questions the accuracy of this statement, and thinks there is not so much difference as might be inferred from it; but he admits there is a difference in favour of the European zamindar.

moderating his rent and so encouraging it. He was told that Rs.3 would be a very fair rent. There are at present some low-lying lands, which are rented at Rs.6 to Rs.7, and even as low as Rs.3, but not in the Berhampur and Surdah districts. In a letter he had received from Monsieur Gallois, of the filature of Messrs. Louis Payen & Co., since he visited the filature, the following opinion is expressed:—

"I told you in the course of our conversation that I consider the decrease of late years in the rearing of silkworms (*Bombyx Mori**) in Bengal to be entirely due to the excessively high rents the zamindars put upon the mulberry lands in years of prosperity in the silk-trade, and which they maintained when the value of silk in the world fell down 20 to 50 per cent.

"I could point out estates in the Maldah, Rajshahye, and Murshidabad districts where rents for mulberry lands are as high as Rs.12, Rs.14, and Rs.16 per bigha.

"I am perfectly aware that the Government is powerless to interfere with the zamindars to bring them down to lower rents for lands which they have collected for many years past, but I believe that a great deal might be done in the estates under the Court of Wards† if the Collectors in

* Not Mori but fortunatus and cræsi.

† NOTE.—It may be of interest to explain to readers out of India the nature and work of the Court of Wards and attached estates in Bengal. Mr. Wardle had been assisted

charge would fix a fair rent for all lands This step would at once increase the mulberry cultivation in these Courts of Wards' estates and be a very good example for the neighbouring zamindars to follow.

"I consider Rs. 3 per bigha a fair rent for the best mulberry lands. The best rice-producing lands are seldom rented higher than Rs. 1-8 to Rs. 2 per bigha."

The industry appears to be crushed because of this high rent. There are hundreds of square miles of land under mulberry cultivation, but the quantity is rapidly declining, and the land used for other crops. The mulberry is not grown as a tree for

in this by an interesting article on the subject in the *Englishman* of the 11th January, 1886.

There are 170 wards, and 90 attached estates under the management of the Revenue authorities, who fix and collect the rents of these estates.

Indifferent harvests have made the collection of rents difficult. In the Bhagalpur Division the collections only amounted to 84·1 per cent., and they were also unsatisfactory in the Burdwan and Rajshahye Districts.

There would probably be increased revenue if the silk industry of Bengal could again flourish, so that more land might be put under mulberry cultivation.

The estates have to bear subscriptions, donations to schools and to dispensaries, and for keeping the property in good order for the benefit of the wards, who are minors, who are educated first in the zilla and other district schools, then in some cases by special tutors, whilst "others are educated at various colleges with private tutors to direct their studies. One good feature in the curriculum of education is that, as soon as any minor reaches the age of eighteen, he is, in addition to his ordinary education, taught the principles of zamindari business, and is practically inducted into the management of his own estate."

silk-rearing in the Bengal district proper, but is planted over the land as a shrub, and does not grow higher than from one to two feet. It is the *Morus indica* and not the *Morus alba* of Italy. Land for mulberry-growing has come to be more highly rented than for any other crop. Even land used for growing rice adjoining the mulberry plots is only rented at twelve annas to one rupee per bigha. A striking difference this it must be owned. Other causes are also in operation increasing the difficulties of recent sericiculture in India, and, as it was one of his missions here to stimulate the industry and to point out reasons for its improvement, he hoped it would not be considered out of place if he placed on record all the hindrances to sericiculture recorded to him in his visits to the filatures. Mr. Morey, of the extensive filatures of Messrs. Robert Watson & Co., of Surdah and other parts of the province, told him that another great reason of the decline of this industry is mainly due to the low price of the silk market and the failure of the cocoon crops. The cultivation of the mulberry, too, had considerably decreased, owing to the reasons already given, and in consequence of the ryots finding better remuneration by sowing rice, sugarcane, and other crops, the cultivation of which had considerably increased. He considers the failure of the cocoon crop to be principally due to the bad seed selected by the native breeders. Formerly they themselves went to purchase seed from other districts, and saw the worms building their cocoons before buying their seed cocoons. In doing this they also rejected any bad cocoons. Now, the system is that brokers, "Sanchoo-Pikars," go instead and buy indiscriminately and sell to the rearers. The consequence of the absence of proper selection is that diseased and weakly worms are bred, the crop fails, and the rearers get disheartened. Mr. Morey suggests that Government should come to the rescue and rear cocoons for *seed* purposes only, and also in growing mulberry properly to feed these seed worms. This latter would take, perhaps, a year to do, but the benefit would be very great; the natives would also gladly purchase good seed. He suggests that at first the Government might distribute the seed to the native rearers gratuitously in order to get them to take it, but as soon as the natives were found to use this better seed eagerly, it could be sold, and funds would be created as well as the silk industry greatly improved.*

Further, Mr. Morey suggests that, as during the hot months worms are killed by the extraordinary hot winds, because of the matted walls used in the construction of the native village huts, it would be well for Government to help in erecting huts with mud walls in all large cocoon-rearing villages, in order to show the rearers that by this plan they

* Mr. Wardle had called upon Dr. King of the Royal Botanic Gardens, Calcutta, who had kindly offered to cultivate the various species of mulberry and to distribute plants gratuitously to any mulberry-growers who cared to apply for them.

would save their worms and secure much larger crops than at present. In those villages where the rearers have mud-walled huts, the cocoons seldom fail. This, he thinks, is a very important fact. Mr. Morey had told him that during the last three years at least 60 per cent. of the worms had died from disease and the hot winds. He thinks Government should employ some responsible person to visit the rearing districts, and show the rearers what is required. He also thinks the zamindars should be asked to interest themselves more. He also thinks it worth suggestion that Government might commence the rearing of brood cocoons in jails under the supervision of a paid expert.

Now, as he, Mr. Wardle, knows from long experience that many gentlemen connected with Government administration in England "pooh-pooh" any request or suggestion for the substantial and practical aid of anything connected with trade or commerce, and are ever ready, too ready he thinks, with their pet expression, "It must be left to private enterprise," it is quite possible that some or all these thoughts may receive similar treatment, yet he cannot help hoping that Mr. Buck, who has repeatedly proved to him the very strong interest he feels in this industry, and at whose call he has at great sacrifice come out to India, will consider the wants of Indian sericiculture with a breadth which he thinks it needs, and a determination not to let the subject pass away until it meets with that great resuscitation which with proper care and assistance awaits it. He earnestly appeals to him to support this industry.

He ventures to say that collective action is what is most needed to restore our silk industry both at home and in India; it is, in his opinion, necessary that Government should take the initiative in a well-thought out application of any lever which will help to raise so beautiful and artistic an industry to its proper and ancient proportions. The wearing down of any artistic industry is a national loss which requires national aid and encouragement.

The silk industry of England has received no real encouragement from her Governments, not even in providing well and correctly compiled statistics. In this respect we are behind every other western silk country, and its present state is dreadful.

As the silk industry gradually travelled westward, Rulers and Governments assiduously welcomed and nurtured its growth, never more so than in France and Italy in the past and especially at the present time. Extraordinary effort is there made outside private enterprise for the retention and extension of this their valued industry. We have an advantage which France and hardly any other nation possesses in having a great silk-growing country in India, capable of supplying not England merely, but all Europe. In England we have silk centres, such as Spitalfields, Coventry, Macclesfield, Congleton, Middleton, and other districts, not long ago great centres of silk industry, now paralysed, and the want of collective action and Government supervision has had not a little to do with it.

In what way it would be best for the Government of India to support this effort is hardly for him further to suggest, but he is sure there is an immensely important future in India, not only for the ordinary silk of commerce, but also for the newer industry of Tussur silk, with which he has had something to do during some years of persevering effort to bring into life. With regard to the silk of commerce, he has been able to convince the gentlemen who have charge of the Bengal filatures he has had the pleasure of visiting, that they are manipulating a fibre quite equal, if not superior, to that of Italy, in respect of structure and regularity. He is certain, from the microscopical and other examinations of many species of silks he has had the honour to make for and at the request of the important Chamber of Commerce at Lyons, a report of which had been circulated in India by the Government, and which should be read and studied at every filature, that Bengal silk is capable of as much refinement as any other silk of the same species, be it Eastern or Western. He repeated that it is all the produce of species closely allied to the *Bombyx Mori* of Europe, and the fibre is, for all practical purposes, nearly the same. With care and perseverance India will realise the truth of these examinations and assertions. On the table will be found specimens of silk which he has seen reeled with the appliances of which he has spoken other than those at present in use. The silk produced is almost as good and pure as Italian, in fact much better than some of the Italian silks.

To those inclined to doubt what he says he commends a careful examination and comparison of the specimens which he saw reeled in the Rajshahye filatures and which now lie on the table.

India only wants the application of progressive observation, and the immediate adoption of whatever method the progress of sericicultural science and mechanical art bring to add better results in Europe. That there is ample scope may be inferred from the fact that for the ten years ending 1883, we in England have been purchasing manufactured silks from countries in Europe to the extent of about eleven millions pounds sterling annually, and that whilst during that period Europe has been annually consuming 6 million lbs. grown in Italy, 1½ million lbs. grown in France, 7 million lbs. grown in China, 3 million lbs. grown in Japan, India can only show an export of one million lbs., and last year it had sunk down to an export of only 457,600 lbs. This does not include that grown for native consumption, which is no doubt very considerable, and has much increased since the very low prices of Bengal silks of late.

He found at Poona, as well as at other places, the silk manufacturers were all buying their raw silk from China on account of superior quality, and had given up the use of the silk of their own country. At Lahore he found Bokhara silk exclusively asked for.

The value of the exports of silk from India in 1884-5 was Rs.46,35,613, whilst that of the silk imported was Rs.74,75,633, or nearly twice her exports.

Is it not high time that collective as well as individual effort should be made to alter so serious a state of things? *

From these statistics it may be inferred that every acre of mulberry land may be multiplied many-fold.

From the short time placed at his disposal by the Royal Commission he had not had an opportunity which he should have desired of visiting those

* NOTE.—He extracts from Mr. J. E. O'Conor's "Review of the Trade of India in 1884-5" the following interesting statement:—

"Silk was imported to a much smaller extent than in the last two previous years, only 1,831,702lbs., which was 17 per cent. less than in 1883-84, though still a very large quantity for a country which is held to be a great silk-producing country. Whatever may be the capacity of the country for producing silk in large quantities, it is clear enough that while India imports more silk than it exports (the bulk of the exports being, moreover, only waste or chassam), the country must more properly be called an importer and consumer than a present producer of raw-silk. Most of the imported silk comes from China and Siam *via* the Straits for Bombay mainly and Burma in smaller degree. Even Bengal, however, the great silk-producing province, imported 212,349lbs. of silk last year. It is understood that this silk, like the silk imported into Bombay, is for the use of a local mill which has been at work for a couple of years weaving silks for the Burma market."

districts in India where sericiculture has been attempted or is being to a less extent carried on, namely, the North-West Provinces, the Punjab, and Kashmir.

The mulberry thrives luxuriantly in these provinces, and he has no doubt there is a good prospect in store for them under sufficiently trained European superintendence and patient development on a moderate scale.

There will be shown at the coming Exhibition in London cocoons of the *Bombyx mori* of excellent quality and form from these provinces.

He would strongly urge that Government should annually ascertain the quantities and values of native-reeled silk used chiefly in India and Burma, and also that produced by European reelers in India for export. In France and Italy the quantity of cocoons annually raised is recorded and published, and it should be so in India; also the quantities of Tussur, Eria, and Muga, not omitting the waste silk of all kinds carefully defined. The want of these statistics is very great, and it is a reflection upon our system of compiling our Board of Trade returns that comparatively no information of value can be as yet obtained. It is particularly necessary, now the Tussur utilizations promise a most important augmentation to the silk trade of India, for if it is properly cared for he predicts for India an enormous expansion of Tussur silk growing and reeling, and, as he is fully convinced that a great future also lies before Bengal in the revival of her at present distressed and unremunerative silk industry, he thinks it will not be denied that his suggestions on this head of accurately compiled returns are of the first importance. He repeats that what is wanted is—

(1) An annual return of the quantity of cocoons raised of the mulberry-fed worm of each bund, distinguishing the univoltine and multivoltine kinds ;

(2) An annual return of silk reeled in Bengal in the European filatures ;

(3) An annual return of native-reeled silk, which could be obtained without much difficulty. All the silk produced in the different districts comes through the bazaars in the raw state, and the Magistrates could easily obtain statistics by finding out the quantities annually sold at the different marts ;

(4) An annual return of all waste, *chassam*, knubs, pierced cocoons produced in the reeling, breeding, and manufacture of silk proper ;

(5) Tussur silk, quantity of cocoons raised in the various provinces ;

(6) Quantity of waste, raw-silk, and cloths of Tussur ;

(7) Similar returns of Eria, Philosamia ricini ;

(8) Similar returns of Muga, Antheræa assama ;

(9) The nature and quantities of the exports of each kind to be severally and accurately defined.

Another subject which had specially occupied his attention was the development of the trade in Tussur silk, and he had been engaged in attempts in this direction for some years past. He could not for some years induce the manufacturers in England to utilise Tussur Silk; he had to visit Germany to have his ideas of Tussur plush first carried out. The Government of India sent home in 1874 a quantity of Tussur cocoons of various sorts, and he had succeeded in reeling them in Italy. Later on people began to see that there was something in Tussur silk. In 1878, at the Paris Exhibition, the French manufacturers were very much struck with his examples of improved reeled Indian Tussur, telling him that they could give orders for almost any amount if it could be reeled as well as that exhibited. He told them that such silk was not yet in the market. He then appealed to India for it, and the Indian producers did not at all respond to this appeal. There was then nothing left for the manufacturers but to go to China and there they have found what they wanted; but although China had got the start in the race in the competition with India, India had now begun at the right end. China began by sending out the worst rubbish she could possibly find, and India had begun by sending her best reeling. The fibre of Indian Tussur silk is a better one than that of Chinese Tussur, which is quite another species, *Antheræa Pernyi.* It must be a great pleasure to all who were connected with sericiculture in India to find several gentlemen vieing with each other to produce the best quality of real Indian Tussur silk; and the task of comparing one production with another had now become a most invidious one. The new appliances and inventions for reeling were all good, but they might all be improved. In an adjoining room will be found various specimens produced by new methods. Mr. Cleghorn had been paying a good deal of attention to the softening and reeling of cocoons, and with a great deal of success. In the silk districts Mr. Wardle had heard high opinions expressed as to the value of Mr. Cleghorn's patent. Mr. Peppé of Arrah had also been working in the same direction, and had produced some beautifully lustrous Tussur. M. Gallois had been working in the same way, and had produced Tussur which was in great demand on account of its softness and excellent reeling. He was supplying extensively the Lyons market, and could not produce it rapidly enough. While at Surdah he happened to be present at the reeling of silk by Mr. Cleghorn's method; Mr. Marshall, the manager of one of the filatures of the Bengal Silk Company, came over to Berhampur to meet him, and brought with him his own developments in the production of Tussur silk. This morning M. Gauthier of Hazaribagh had brought examples of his reeling. They were all beautiful, and the gentlemen present could judge which of them were the best. Specimens of each kind of improved reeling lay on the table. Mr. Wardle urged the adoption of the *Tavelette*

Consono in the reeling of Tussur cocoons. He believed it to be indispensable and as yet the best and most simple appliance for reeling. Fig. 5, plate 19, is a full-sized drawing of this little instrument. Its position in the reeling machine is easily recognised in fig. 6, plate 20. The framework is of brass.

A. A. fig. 5, plate 19, enamel guiders through which the combined threads pass from the reeling basin and *filière* upwards to the reel.

B. B. light pulleys of wood with glass spokes revolving on steel pins.

C. the *croissure* or crossing of the silk whereby a gentle friction upon itself by itself is given. This method of *croissure* is superior to that which was used in France and Italy some years ago and known as the *Tavelette Chambon*, and which he had described and figured in his "Handbook of the collection illustrative of the Wild Silks of India in the Indian section of the South Kensington Museum" in 1881. It has since given way to the improvement effected by Consono in the form of the *Tavelette* known by his name and illustrated in fig. 5, plate 19. Of its superiority and efficiency too high praise cannot be accorded. For reeling Tussur cocoons it is absolutely necessary as he had proved in the several filatures where he had mounted and worked it.

D. point of attachment by a screw to the Tavelette stand or support.

The difficulty of softening the cocoons preparatory

PLATE 19.

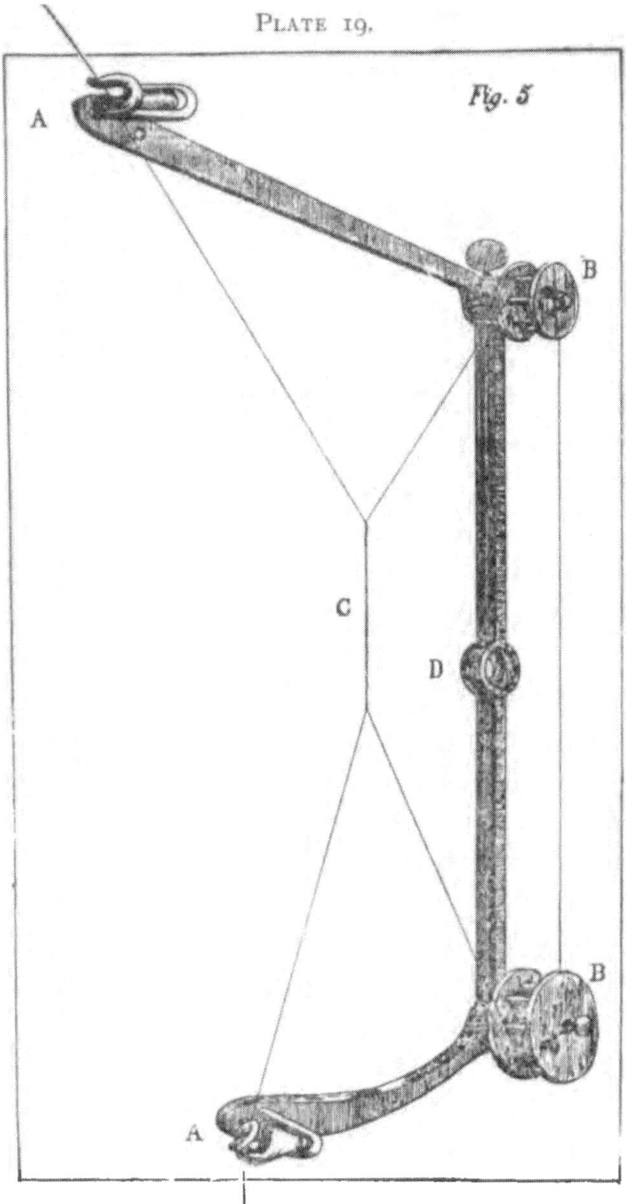

Tavellete Consono.

to reeling is evidently wholly overcome in each of the new methods. Submitting them to the action of steaming appears to him to be the chief point in overcoming the difficulty of softening* so hard and shell-like a cocoon. It was now left to the Government to come forward and facilitate the cultivation of Tussur silk and the collection of cocoons, as well as giving facilities for buyers of cocoons to transmit cash readily into the innermost jungles to pay the natives on the spot. It is at present most difficult to send money into the interior. Tussur was entirely wild and uncultivated, and would possibly so remain, but it could be and was managed as an outdoor production. These results had clearly proved that Tussur silk could be well reeled and produced and made to pay as an article of commerce, and Mr. Wardle considered his work in connection with it was coming to an end. He had been assured that Tussur silk as an article of dress was rapidly gaining favour and becoming very fashionable in France, and that the demand for it might increase to such an extent as to make its price almost as high as that of other silks.

There were two other silks which now engaged his attention, and which would be found mentioned in his Handbook published in 1881 and the various

* Since the meeting he had visited Futwah in the Behar district and had seen the natives reeling Tussur and found that they softened their cocoons perfectly by boiling them for three hours with *sagi* and reeled them by hand and thigh, a decoction of myrabolans being used as a lubricant and to cement the thread.

Government Reports before and after that time. Since then he had been working at them, and he could now speak of them with much greater confidence. The first was the Eria silk of Assam, the worm of which fed on the castor-oil plant, a plant which he found abundantly all over the district he had just visited and in the other parts of India he had travelled through. He suggested that they could as well cultivate the Eria with the mulberry-fed worm at the same time, and he thought this would soon be tried. The cocoons of this worm could not be reeled to pay, and he would suggest they be packed up in pressed bales and sent to the spinners of Europe to be carded and spun. The next was the Muga silk produced by the larvæ of *Antheræa assama* at present chiefly an Assam worm but found also in North-East Bengal feeding on the Soom tree, *Machilus Odoratissima*, a laurel which abounds in Assam.

Of this silk he could not report too highly; it had an immense future before it. It was already exported from Assam to Calcutta, and beautiful embroidery made with it. At present the only difficulty in connection with it lay in its limited production, in overcoming which lay bright hopes of promise, for the improvements effected in Tussur reeling applied exactly to Muga. He should have more hopes for it than for the other. It was a beautiful silk, which took dyes more readily than Tussur, and was worth the attention of every one who was interested in producing it.

The immense district of Chota Nagpore was being opened up to Tussur enterprise. Cocoons are found abundantly, and agents at Ranchi are now buying them largely from the natives, who bring them in from the jungles on the plateau, 2,200 feet high, where he believes both the mulberry and Soom trees would flourish, and the *Bombyx Mori* and Muga silk, if not Eria, could be cultivated advantageously, because labour is there so cheap and plentiful. Besides this district, Mr. H. G. Turner, Collector of Vizagapatam, has just arrived in Calcutta and told him of an enormous area there where Tussur occurs. The other day a firm in Madras purchased Rs.42,000 worth of Tussur cocoons from the Chota Nagpore district.

Mr. Wardle then explained the method of working the reeling-machine which is universally in use in Bengal, and then contrasted it with the improved reeling-machines used in the largest filatures in Italy. Fig. 6, plate 20, shows the whole machine at work and the following is a description of its parts :—

A. Woman reeling cocoons.

B. Construction of brick supporting the bassine with fire or steam arrangement as desired.

C. Bassine from which the cocoons are reeled.

D. Four cocoons being reeled together.

E. Small bassine, in which the cocoons are first washed, softened, and arranged for the reeler.

F. Filière uniting the four cocoon baves into one thread of raw-silk.

G. The Tavelette Consono.

H. The tavelette support.

I. Brush by which the reelable threads of the cocoons are found. It is turned downwards when used, and is worked backward and forward.

J. Bassine for another reeler.

K. Pulley guiding the silk to the reel.

L. Reel on which the silk is wound.

M. Pulley and strap for turning the reel. The turning can also be done by hand when desired for cottage reeling. The silk when taken off the reel constitutes the raw-silk of commerce.

Mr. Wardle stated that the value of Italian silk in England was about 19 shillings per pound; whereas the value of Indian silk in England, reeled by the crude apparatus in use, was but from 8s. 6d. to 11s. 6d. This difference was mainly due to the different method of reeling and to the great attention paid to the sizing of the silk in Italy. He also explained the difference between the kinds of labour employed in the two countries and the greater speed at which the reels were driven in India compared with that of Italy. He concluded by stating that he thought he had sufficiently explained what he had to say, and had exhausted his subject. He was much obliged to all present for the earnest attention which they had paid to his remarks. He was obliged to all those gentlemen who had aided him in his visit to the silk districts, and particularly to the kindness of Mr. Keswick and Mr. and Mrs. Morey of Surdah.

PLATE 20.

Woman reeling Cocoons.

Mr. J. J. J. Keswick then proposed a vote of thanks to Mr. Wardle for the very interesting report on what he had seen and heard, and he felt certain that the suggestions he had been able to give to the managers of silk filatures in Bengal would produce most satisfactory results.

Mr. Buck said he had one word to add to what had fallen from Mr. Wardle as to the assistance which was expected from the Government. The policy of Government was to confine any assistance which it offered to private enterprise to the pioneering stage, and that in most directions silk enterprise had passed that stage. He admitted that Government might perhaps do something towards investigating the various kinds of silk-worms and moths of the country, and in ascertaining whether the area of useful worms could in any case be extended to other parts of India. The recent investigations in connection with the Exhibition have added much to our knowledge of the distribution of silkworms, &c., and that there seems no reason why some of the best wild silks should not be more widely cultivated. The Bengal Silk Committee have been invited to place before the Government any suggestions they might be able to make as to the directions in which Government aid would be otherwise useful.

Mr. Keswick remarked that it might be worth the while of the Government to take an interest in improving the *Bombyx* cocoons of Bengal, but he was not prepared to say how they could best assist private enterprise in this direction. He did not believe in importing seed but in the more skilful treatment and selection of what we had in the country.

Mr. Liotard remarked that the real objection to the cultivation of the silkworm lay in the high rental of land. There was one class of cultivators of mulberry, who sold the leaves of the plant, and the rearer of the worm, who purchased the leaves, often, to raise his income as much as he could, did not feed the worms sufficiently, which resulted in their producing a small quantity of silk. There was no hope for improvement, unless the zamindars reduced their rents.

Mr. Keswick thought that land, like anything else, must be governed by the laws of supply and demand.

Mr. Buck thought that it might be safely left in the hands of the Bengal Committee to consider any steps which it might consider useful to take in the matter.

Mr. Wardle, in acknowledging the vote of thanks, wished to call attention to the interesting entomological and industrial additions which had that morning been made to the collection he had brought to India illustrative of the utilisations of Tussur silk in England, and of the examples of the successful application of the tinctorial matters of India, paying a strong tribute to the value of Dr. Watt's most interesting collection of raw products in Indian cocoons, silks and dyeing plants, and of his energetic collection of them throughout India, and also to the great benefit which would accrue to Indian sericiculture by Mr. Buck's wisely calling in the aid of the resources of the Calcutta Indian Museum with its beautiful entomological collection now in the next room, and to the scientific investigations of Mr. Wood-Mason, Officiating Superintendent of the Museum, who, at the request of Mr. Buck, had undertaken to prosecute a thoroughly scientific and continuous examination of the numerous silk-producing larvæ and their moths of India, a more complete knowledge of which is anxiously called for in both England and France, and which will greatly aid the silk rearers of India in lessening disease and in promoting goods bunds.

A vote of thanks to the Chairman brought the meeting to a close.

Plates 18 and 20 were drawn at the Mayo School of Art, Lahore, under the superintendence of J. L. Kipling, Esq., Principal, during my stay there as Mr. Kipling's guest, by Shah Muhammad, chief student. Fig. 5 plate 19 was drawn at Calcutta.

CHAPTER XXII.

SPORT IN BENGAL IN 1886. BIRDS. FOXES, JACKALS, CROCODILES, MY TIGER HUNT, ETC.

IN this final chapter I hope I may not unfitly supplement the account of my Sericicultural work in Bengal during my first visit to India in 1885-6 as recorded in the preceding chapter, and narrate the occurrences of a few of the happy days of sport I was privileged to enjoy with the rifle and shot-gun, and on horseback, especially as I had the exceptionally good luck to kill a tiger, the account of which I wrote at the time, but which has not been previously published.

On learning from me that I greatly desired an opportunity to shoot a tiger, Mr. J. J. J. Keswick, of the firm of Messrs. Jardine, Skinner and Co., the zamindars of the property in which the Surdah and other filatures are situated, telegraphed from Calcutta to Surdah to enquire if any tigers had been seen in that neighbourhood, and on learning that two had been lately seen, I left Calcutta on the night of the 27th December, 1885, by the Eastern Bengal State Railway to visit the principal silk filatures of the Rajshaye and Murshidabad districts in the interests of Indian Sericiculture at the request of the Government of India. Four o'clock next morning brought me to Damookdiaghat on the Ganges which I crossed in a steam-ferry, and proceeded by train to Melanchi station, on the Darjeeling railway, where a car was waiting for me from Surdah, some fifteen miles by road through jungle of bamboo and mango topes.

At Galampore is the first cocoon-reeling branch of a number of country filatures in the Rajshaye district. I stayed at this filature on my way to Surdah to get my first experience of Indian cocoon-reeling. The sight of the golden yellow cocoons as they lay occupying a large space on the ground, drying in the sun, was very striking.

I breakfasted with Mr. Renwick, who has an extensive business in this district in supplying it with roller sugar-cane crushers. Sugar-cane cultivation is a large industry in this part of India, and the growing, cutting and crushing of the cane are very interesting operations.

I arrived at Surdah in time for tiffin, and the rest of the day was spent in the inspection of this extensive filature, the most important in Bengal. It is situated by the Ganges, close to the house occupied by Mr. and Mrs. Morey, my host and hostess. Mr. Morey is the director of the silk industry of the district.

At sunset I had a stroll with my gun and

killed a few interesting birds, also a jackal which I could not then find owing to darkness coming on so quickly at night. A great number of them came prowling about the filature in search of the refuse chrysalides of which they are very fond. The howling they make during the night and early morning is appalling. I think all the jackals of the neighbourhood came under the window of my bedroom as if to avenge themselves for the slaughter of their comrade. They howl and shriek to each other in batches from river side to jungle. Their howling is humourously imitated locally by "Here's another roast Hindoo." "Oh! where, where, where?" "Oh! here, here, here!!"

A welcome dinner at eight, and bed at ten ended a pleasant day.

Next morning before seven Mr. Morey gave me a mount, and we trotted off with two resident friends to the Maidan for a gallop, one of whom was Mr. Walton, now Director of the Kashmir filatures; we had four dogs of various breeds, accustomed to the chase. We soon put up a hare, and had an eleven minutes run; then a fox, which gave us a splendid run, then two runs after jackal and another after a second fox. We were unlucky in every case, the dogs not being in good form and the animals managing to escape either into jungle or sugar-cane bush. We returned at nine to breakfast and spent the day at silk investigation. In the evening I shot a kite, then found my jackal of the previous evening with a pair of vultures upon him, and several grey headed crows. They had skinned him for me beautifully, and had left nothing but the skin and bones. As I approached, the vultures easily got on wing and I killed them right and left. I shot also a beautiful yellow mango-bird, a very large horned owl, and two species of swallow-tailed gulls on the Ganges banks. The indigo plant is extensively grown in the mud flats left on the banks and islands after the rainy season. It is an enormous industry in this part.

Next morning we were in our saddles again at sunrise, in the fresh and cool air, and we took spears to try for pigs, traces of which we had seen the previous morning. We rode out several miles, and our first chase was a fox which after a fast run we killed, two of us being in at the death. We then took a wide beat through green crops in another direction and found again, and after a most exciting run of three quarters of an hour again killed, I alone being there to see. My host, an excellent rider, in trying to turn the fox had been unhorsed, but not much hurt, by a bough lifting him out of his saddle. We came upon plots of sugar-cane into which the pigs had been seen to retire two days before, and with the help of forty natives, sugar-cane cutters (with just enough clothing but not too much), began to beat the canes. It was a fine sight to see the coffee-coloured skins and fine black hair of these villagers as they beat with loud shouts the sugar plots.

We were unsuccessful, and hurried homeward, having a good run with a jackal which I very nearly speared as he ran into his beloved sugar-cane cover, eight to ten feet high, and from which no efforts could dislodge him. It was ten when we returned, after a heavy but delightful morning's hunting. The open spaces of country are of great extent, quite flat between jungles, but the ground is rough with deep ditches and low raised field divisions. In the afternoon we rode out to a silk-reeling village where native reelers were working outside their mud huts. The villages are almost hidden from the sun by large trees of cocoa-nut, date, jan palms, banyan, peepul, tree-ferns, plaintain, with its beautiful broad leaves and fruit, mango and bamboo topes (with the slender tops of the latter gracefully rising over and bending above the other trees), the cocoanut palm, also the Bel (Aegle Marmotis,) the sacred tree of India from which they make their sherbet, with its bare stemless trunk and fanlike top, standing out high above the other trees, a striking sight in the beautiful golden evening after-glow.

The native reeling is very badly done, and the silk not fit for export. It is used for dhoties, chuddars and saris for women's and men's wraps.

It is a curious sight to see the villagers and their sericiculture completely hidden from the ordinary traveller, who sees from the road nothing but forest jungle.

In the evening I shot two jackals, a large Brahiminee duck, and a blue jay.

The charm of Surdah is Mrs. Morey's beautiful lawn and extensive garden, a grand floral sight. I never saw such a profusion of flowers. The hibiscus, bouganvillia, oleander, ponciana with its immense pods, gold-mohr tree, Livingstone and other palms, great and striking kinds of coles, begonia, the grand red poinsettia in long hedges, also white ones, tree-ferns, cones of immense convolvuli, African marigolds, large cockscombs, a large slender shrub a mass of orange flowers, and many others, made up such a floral scene as I never saw before, the whole bounded on the east by immense casuarina trees—landmarks for miles by their great height, and on the west by a great avenue in which the banyan and peepul trees (always planted together as man and wife) were conspicuous by their great size.

Mr. Morey had arranged for a couple of days shooting in the jungles at Rhurclikia, and next morning we set out at eleven, and drove through eighteen miles of delightful jungle, passing the English station of Rampur Bauleah in which are located the Judge, Magistrate, Doctor, and other officers charged with the administration of the needs of the locality.

We arrived at a long stretch of short dry grass about a mile from Rhurclika, where we were met by twenty coolies waiting to beat for us. We took off our coats and waistcoats and began to beat for partridges. It was scorching hot, and very trying in so bright a sun. We killed two brace of black partridges which afterwards proved excellent eating, a large quail and a wee button quail about the size of a jack snipe, and a beautiful large white hawk that swooped down upon the little quail the moment it fell. It must have been following us, falcon like, high above waiting its opportuuity to swoop down upon any game we killed. Nothing lies many minutes undisturbed nor remains in *corpore vili* long; vultures by scores, crows, eagles, hawks, kites by hundreds, jackals, the white ant and other scavengers soon leaving nothing but the bare bones of any bird or animal, be it ever so big.

After shooting a few other birds and a hare, we mounted ponies and rode to Rhurclika, an indigo factory on the Ganges, with a house beautifully situated on a bend of the river, here about two miles wide. We had a most glorious sunset over the water, and the bend of the river is such that the sun both rises and sets over its waters.

After dinner I strolled out in the starlight to

PLATE 21.

Preparing for the Tiger Drive. The Howdah fixed.

Plate 22.

Waiting to commence the Tiger Hunt in the Maidan Bamboo Grass in Bengal.

look at the elephants that were to carry us on the morrow. Several had arrived and were crunching up leaves and branches of trees, and near them were two clusters of natives, mahouts and coolies, sitting round their fires telling stories of former hunting adventures, and no man has a greater right to recount his dangers than a mahout, as sitting on an elephant's neck he is often the first to be pulled off when a tiger charges, his is always the post of danger, and no nervous or fearful man could take it.

We were up by sunrise the next morning, and I went out to see the howdahs put on the elephants. (Plate 21.) The roping of the howdah to the elephant's back with its big mattress between is a rough operation, and it is well the animal is pachydermatous. Two of the elephants were bathing in the large tank, their trunks upraised, and were swimming about with ease, enjoying their bath immensely. They came out at the call of their mahouts, and were soon dressed up for the chase waiting to start. (Plate 22.) We set out for the jungle—a most goodly sight—the elephants and forty half naked coolies with sticks, one carrying the ladder for mounting and dismounting the elephants, others with necessary odds and ends of things that would not bear the shaking of an elephant's back. We soon met a dusky farmer, who, putting his hands together, besought us piteously to come to his farm and kill a leopard which had during the night carried off a calf for which he had only paid fourteen rupees the night before. He led us through tangled jungle until we came to his little holding. He showed the broken enclosure out of which the leopard had dragged his calf, and we tracked it through two green fields of young crops, of which at least half a yard wide had been destroyed; and under some thick bushes we found the remains of the poor calf. We set about beating for his slayer, but after two hours of unavailing search we gave it up and went deeper into the jungle to try to make the acquaintance of the Royal Bengal Tiger.

After passing through and beating some thick jungle, we came upon a large open plain or maidan, about two miles long and half a mile broad, which during the rainy season, by the overflow of the Ganges becomes a lake, and which afterwards dries up and a large growth of high bamboo grass succeeds. We commenced to beat this maidan in line, with beaters both between and flanking the elephants; we beat the whole of the ground except a small portion, disturbing nothing but partridges and a few jackals and foxes. We saw abundant marks of wild pigs and several large circular tiger lairs, in one of them the remains of a cow killed not more than two or three days before. We returned to beat the thicker jungle by the side of a jheel, probably an ancient course of the Ganges, about two miles long and a hundred yards across, with here and there a little island, on one of which were three crocodiles basking in the sun, a beautiful and impressive sight in the solitariness of deep jungle. Our search for tiger prevented our running the risk of disturbing them by firing, so we merely paused to watch them leisurely glide into the water and disappear. We beat the jungle for the rest of the day unsuccessfully but saw several footprints of the tiger, and set out with a "parvanah" (never mind, as the natives say) from my friend, for our night's resting place. Mr. Morey had ordered his boat to meet us two miles higher up the Ganges than our previous night's resting place, in order to be nearer to the next day's beat.

We retired to new lodgings on board a capacious boat, to dine and sleep on board. Next morning we were out at sunrise in the bracing fresh morning air, and by 7-30 were all on our way again to the jungle. As there was much uncertainty as to finding the tiger, my friend suggested that we should beat for partridges, but he allowed me to overrule him, and we stuck to the tiger quest.

We beat fruitlessly over a considerable range of jungle, and decided to try a more thorough beat in the high bamboo grass of yesterday's maidan. We accordingly beat it up and down in wide beats, leaving no part unvisited. The bamboo grass was generally from about five to seven feet high, and in many places not less than eight to ten feet; splendid cover but not difficult for the beaters to get through.

We were rewarded at length for our trouble, for on getting into the only part we had neglected beating yesterday, the elephants became excited and gave unmistakable signs that something was near; presently someone called out "Tiger" and there he surely was bounding across our front, 150 yards away, Morey and I both fired, the tiger stopped momentarily, and then went off at a leisurely trot. He then tried to outflank us, not wishing to leave the high grass. For some time we had great difficulty in seeing him as he often stayed in hiding. We outflanked him several times, and at length drove him to some rising ground between us and the jungle, where he stopped and turned to look at us. I fired at him about 250 yards away, and my bullet went just over his shoulder and struck the ground beyond him. We presently got nearer and I fired at about 150 yards and hit him in the rear. He seemed very undecided whether to charge us or not; after taking stock of us and the situation, he decided to proceed and haltingly disappeared into thick jungle. We put our elephants, now very excited, into a trot, but for some time could not find any trace of him. We had given him up for lost, and on returning by a slight detour we found him under a dense mass of wild plum bush, zizyphus jujuba—one of the trees on which the Tussur silk-worm feeds.

After several ineffectual attempts to get two of the smaller elephants to oust him, we sent in the big one, said to be a most savage brute and fond of fighting a tiger.

He did not like the work, and it was amusing to see him as assiduous in piling up scrub to place a barrier between him and the tiger, as he was at first in tearing it away; his mahout forced him in, but after a terrible charge of the tiger right on his forehead, he retreated with force enough to have knocked down a house had it been in his way.

The roaring and screaming of tiger and elephants was awful; the ground seemed to shake. After a brief struggle the elephant shook off the tiger who retreated into the bush. It was of no use trying to get him again anywhere near the tiger, he had been hurt and much scratched in the encounter and had had enough of it. We were close to him when the tiger charged, but could not see to shoot for the bushes, and it was well the elephant did not back upon us for we should certainly have been upset. There was nothing left but to put our elephant at the high bushes and see what would come of it. We advanced a few paces, very cautiously, what with coaxing and hard hitting, our mahout got him up to the thicket enclosing several big trees where the tiger lay completely hidden. He very cautiously lifted aside with his trunk a portion of the plum bush, just enough to show us the tiger within three or four yards of us, growling and prepared to charge us. He had not time, for the moment the view of him came, Morey called out, "Now," and a shot from my Holland's single barrel Express, a splendid weapon, laid him low. It was all over in a shorter time than it takes to write it.

The coolies, who had run away, slowly gathered round from their hiding-places, but for some time none of them dared venture near the tiger, who lay hidden by the falling of the curtain of boughs. The shikari came up and peering round the beast announced a kill. I then got down by the aid of the ladder which the ladder-bearer cautiously brought up, and with a coolie crept under the plum bushes to the tiger. He was quite dead; I tied a rope round his neck and shoulder and it took about a dozen of us to drag him out. He was a splendid beast, measuring ten feet three inches from nose to tip of tail.

As soon as he was well laid out in the open, natives who had flocked in, and our coolies, began to cheer, and for a few minutes we all had a good bout of it. It was a very picturesque sight, nearly a hundred almost naked brown skins cheering round

After the Kill. The Tiger attached.

the tiger, the elephants looking on very much relieved and satisfied. We mounted and roped the tiger across the back of one of the elephants (see plate 23) and after sandwiches and welcome bottles of Bass, a most enjoyable lunch, we set off to search for a tigress which the natives told us had been seen half a mile further in the jungle, and near to the jheel of yesterday. It was too thick to be beaten thoroughly, and as it was now three o'clock, we decided to go up the jheel where we had seen the crocodiles. We soon came upon a small one about eight feet long basking on the slope below us. We timed our shots and rolled him over in the water. He struggled for a few minutes and then sunk in deep water.

We came to the little island where we saw three yesterday, a large one at least 15 feet to 16 feet lay on it. We again timed our shots at its head and shoulders with such effect that the contortions of its long heavy tail were awful to see. He wriggled himself into the water, and came up in a little while on the opposite side, another couple of timed shots laid him out, but he slid down the waterside into the deep, and as it was dangerous to get off the elephants, and a long way round to the other side, we were unfortunately obliged to leave both of them.* The

* A few weeks afterwards I had the good fortune to kill six crocodiles in one day from a small boat floating down the Ganges a few miles below Cawnpore.

crocodiles were of the species *crocodilius palustris.* A long ride home brought us by six o'clock to Raleetollaghat where our boat was moored. On our way to the river we shot a fish eagle, a beautiful large wild lynx and several interesting birds.

We dined on board and at half past seven o'clock I took my leave of my good friend, Mr. Morey, in a small rickety bamboo boat with two local boatmen, also sixteen coolies who had to carry me in palky next morning to Berhampore, and who squatted with me under a cover of cane work about three feet above deck. In this hole I was rowed and towed about the Ganges until midnight. I had only to cross the river one and a half miles as the crow flies, and to land two miles lower down, and should only have occupied about two hours in crossing, but there were long mud banks which had to be passed, on which we often came aground amidst loud vociferations of alarm from the coolies. At last we got to the other side, and a man got out and towed us with a thin line, which broke and away we went drifting. It was the most comfortless and anxious night I ever experienced, very cold, just starlight enough to see where we were, amongst an oasis of mud banks split up into chasms by the rains.

At last we reached a landing-stage, but to my disgust the boatman refused to land us, and after a long harangue with my coolies and they with me in our own respective languages, he jumped into the water and waded up to his middle to the shore, having fastened his boat a few yards out of reach of landing, and disappeared. The coolies began talking to me but not a soul could understand or speak English nor I their language. It was clear, however, that he had bolted to leave us there until backsheesh liberated us. After waiting nearly half an hour, I began to be very cross with my sixteen stupid men, but they were quite helpless, so I unstrapped my revolver and took the lamp, and put my arms on the shoulders of two men who carried me to the land. I then went in search of the boatman, whom I found crouching under a wall some distance away. I showed him the revolver and pointed to the boat, used mild English language and made him go back and land both the men, the palky, and the luggage.

I then unstrapped my bedding and arranged it in the "palky" (palanquin), lay down inside as well as I could in five feet of room, and was hoisted on the shoulders of four of the men, and consigned myself to fate and sleep. The men began a sort of doleful, melancholy kind of chant, rather Gregorian, and continued it nearly all the way to Berhampore, and I found it was a much easier mode of progression than in a howdah, the sixteen men taking turns at carrying me.

On my friend Mr. Morey's visit to England some time after I told him of this circumstance and asked him if he knew what the words the men were singing were likely to be. He said very probably it would mean

"The Sahib is a heavy man,
A very, very heavy man."

I felt sorry, the explanation was a blow, my imagination had led me astray.

It was a curious palky experience altogether from Ralleetollahghat to Berhampore, thirty miles in all of broken sleep. I awoke before daybreak, and found approaching daylight a blessing.

I had only brought a little bread and cheese and whiskey and water. I arrived at Berhampore at eleven o'clock, and after a good breakfast, I occupied myself in visiting the filatures of the Bengal Silk Company, and that of Messrs. Louis Payen and Co., at both of which I was most kindly and hospitably received, at the former by Mr. Stocks, and at the latter by Monsieur Gallois.

The tiger was skinned the next day at Surdah, the skin was sent down to Calcutta to be cured en route for England to Mr. Rowland Ward, who set it up in the attitude in which it was shot, and it now adorns my shooting box at Swainsley.

APPENDIX.

The following is a list of Silk Filatures and Mills in Bengal.

It has been compiled for me by Mr. B. J. Rose, the head of the India Trade Enquiry Office, Portland House, 73, Basinghall Street, London, E.C., from Thacker's Indian Directory for 1904.

It will be seen from this extensive list how important and progressive Sericiculture is in Bengal.

There is no better silk produced for sewing-silk purposes, it takes all dyes beautifully and with excellent lustre; it is also much used in the manufacture of crêpe, crêpe-de-chine, and similar light textiles, as well also for the silk textile requirements of India, and for many other textile purposes.

It is one of those important branches of Indian Industry which forcibly emphasises the great desirability of obtaining from our own Empire all raw material, which, with its varying climates, it is possible to produce.

LIST OF BENGAL SILK FILATURES.

BARAGARIA SILK CONCERN.

Filatures—Baragaria, Chula, and Jalalpur.

Bengal Silk Co., Ltd.

Gonatea Division: Filatures—Gonatea, Somergunge, Kattasore. Berhampur Division: Filatures — Rangamatty, Chiretty, Bilser, Azimgunj, Joykristapur, Sucktipur, Kamaberi, Beldanga, Babulbona, Banjetty, Bhudderpur and Rajtha.

Berhampur Silk Concern.

Filatures—Rangametty, Chiretty, Geridaripur, Binker, Joykristopur, Sucktipur, Kamarberi, Beldanga, Beldanga No. 2, Banjetty, Rampara, Bhudderpur and Koyta.

Dharupur Factory.

Rearing Stations—Dharupur, Rampur, Agai, Lalganj, Bhatin, Bhadari, Purehasi.

Dun Sericiculture.

Filatures—Lister Grant and Dehra Dun.

Furridpur Silk Concern.

Filatures—Furridpur, Cattamari and Sankarpur.

Kalakankar Factory.

Rearing Stations—Kalakankar, Hinahun, Gajapur, Paingon, Janmamin, Rajinapur and Bazidpur.

L. Payen & Co.'s Silk Filatures.

Cassella, Khajapur, and Shahebgunge in Rajshahi, Bholahat in Maldah, Plassey in Nudden, Goorellee in Midnapur, and Gadi, Choa, Teakattah, Dharrumpur, Narainpur, Bazarpara, Panchkattiah, Suzapur, Kapasdanga, Hatibanda, Satoye and Gouripur in Murshidabad.

SARDA (SURDAH) SILK CONCERN.

Sarda, Motihar, Dacra, Chattaria, Pananagur, Doorgapur, Mirzapur, Belgaria, Bharraghurria and Checla, and also at Furridpur.

ROSE SILK FILATURES.

Ramnuggur, Beldanga, Balligotta, Shaitora, Shirool, Babda, and other places in Rajshai.

INDIAN FILATURES AND SILK MILLS.

	FILATURES.		MILLS.	
Year.	*No. of Works.*	*Persons Employed.*	*No. of Works.*	*Persons Employed.*
1893.	73	8,638	28	3,045
1894.	78	9,360	27	3,143
1895.	89	9,018	28	3,586
1896.	68	8,263	29	3,914
1897.	67	5,625	31	2,795
1898.	66	13,918	31	2,773
1899.	67	14,139	33	3,092
1900.	66	12,292	29	3,220
1901.	71	11,061	12	3,102
1902.	62	10,652	10	3,040

FINIS.

PLATE 24.

Kashmir Woman spinning wool.

PLATE 25.

Kashmir Women husking and winnowing rice.

Srinagar Hindoo Woman carrying water.

A Jhelum Steering Boat-girl, with the customary heart-shaped oar.

PLATE 28.

Vegetable Bazaar in one of the streets in Srinagar.

PLATE 38.

Sweet-makers' and sellers' Bazaar; Street in Srinagar.

PLATE 39.

Page 375.

The Potter's Wheel and Work in a Kashmir Village.

PLATE 43. Page 376.

Fakir or Holy Man on Lunka Island, Dal Lake, sitting in the snow.

PLATE 44.

View of the terraced agricultural operations between Srinagar and Islamabad